Hemjyoti Kalita

Shape Memory Polymers

Also of interest

Polymeric and Natural Composites
Materials, Manufacturing, Engineering
Bououdina, Davim (Eds.), 2018
ISBN 978-3-11-049458-7, e-ISBN 978-3-11-049317-7

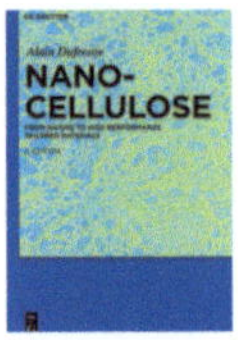

Nanocellulose
From Nature to High Performance Tailored Materials
Dufresne, 2017
ISBN 978-3-11-047848-8, e-ISBN 978-3-11-048041-2

Polymer Engineering
Tylkowski, Wieszczycka, Jastrzab (Eds.), 2017
ISBN 978-3-11-046828-1, e-ISBN 978-3-11-046974-5

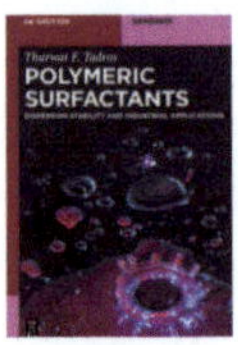

Polymeric Surfactants
Dispersion Stability and Industrial Applications
Tadros, 2017
ISBN 978-3-11-048722-0, e-ISBN 978-3-11-048728-2

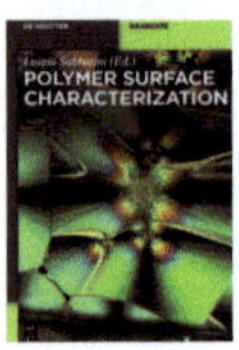

Polymer Surface Characterization
Sabbatini, 2014
ISBN 978-3-11-027508-7, e-ISBN 978-3-11-028811-7

Hemjyoti Kalita

Shape Memory Polymers

Theory and Application

DE GRUYTER

Author
Dr. Hemjyoti Kalita
Laboratory for Advanced Research
in Polymeric Materials (LARPM)
Bhubaneswar 751024
India
hemjyoti123@gmail.com

ISBN 978-3-11-056932-2
e-ISBN (PDF) 978-3-11-057017-5
e-ISBN (EPUB) 978-3-11-056941-4

Library of Congress Control Number: 2018934739

Bibliographic Information Published by the Deutsche Nationalbibliothek
The Deutsche Nationalbibliothek lists this publication in the Deutsche Nationalbibliografie; detailed bibliographic data are available on the Internet at http://dnb.dnb.de.

Cover image: Power and Syred/Science Photo Library
Typesetting: Integra Software Services Pvt. Ltd.
Printing and binding: CPI books GmbH, Leck
♾ Printed on acid-free paper
Printed in Germany

www.degruyter.com

Preface

Shape memory polymers are an emerging class of intelligent polymers, which are gaining tremendous interest due to the possibility of potential applications. Shape memory polymers can be deformed and temporarily fixed in a second, temporary shape. This temporary shape is retained until the shaped body is brought to a suitable stimulus, which actuated the recovery of the original shape. In this book, the basic principles and mechanism of shape memory polymers, classification of shape memory polymers and related characterization techniques are illustrated. Furthermore, an overview of the broad spectrum of applications in various fields for shape memory polymer is presented. The first chapter comprises basic principle and mechanism of shape memory polymers along with the brief descriptions of the hyperbranched shape memory polymers, blend shape memory polymers, interpenetrating shape memory polymers, biobased shape memory polymers and shape memory polymer nanocomposites. Second chapter consists of classification of shape memory polymers based on nature of structure and actuation techniques. In the third chapter, various characterization techniques and properties of shape memory polymers are presented. This chapter gives the detailed discussion of the effect of shape memory characterization techniques on the shape memory properties of shape memory polymers. Fourth chapter gives the brief description of materials, synthesis and new possibilities of shape memory polyurethanes. This chapter also focuses on the effect of material structures and synthesis techniques on the properties of shape memory polyurethanes. Fifth chapter deals with the applications of shape memory polymers in various fields and their new emerging field of applications. The purpose of the book is to cover the theory and comprehensive discussion to recent developments in the shape memory polymer along with their possible field of applications. I hope that readers will appreciate the choice of the topics included in the book.

Hemjyoti Kalita

https://doi.org/10.1515/9783110570175-201

Preface

[illegible]

Contents

1 Shape memory polymers

1.1 Introduction

Shape memory polymers (SMPs) are stimuli-responsive smart materials that can respond dynamically to the environmental stimuli [1–4]. Nature shows numerous examples of stimuli-responsive phenomenon. For example, sunflowers bend toward the sun to maximize light exposure, the leaves of *Mimosa pudica* fold inward quickly when they are touched, and chameleons change color according to the environmental situation [1, 5]. Thus, to mimic nature, huge efforts have been emphasized to design stimuli-responsive polymers. The first SMP, a poly-(norbornene)-based polymer (T_g = 35–40 °C), was invented in 1984 by CDF Chimie Company, France [6, 7]. Two more SMPs were also invented soon after it; one was poly(trans-isoprene) (T_g = −68 °C), which was invented in 1987 by Kurare Corporation, Japan, and another one was poly(styrene-butadiene) (T_g = 60–90 °C) by Asahi Company, Japan [8]. However, all these SMPs suffered the same problem with limited processability. Along this time frame, thermoplastic polyurethane (TPU) SMP was invented by Mitsubishi Heavy Industries Ltd., Japan [9]. It stimulated the significant interest in SMP, presumably due to the versatility of urethane chemistry with wide range of T_g and easy processability. Currently, Cornerstone Research Group, USA, has developed and commercialized polystyrene-based SMP (Veriflex®, Verilyte™ and Veritex™); SMP Technologies Inc., Japan, thermoplastic and thermoset polyurethane-based SMP (DiAPLEX™); Composite Technology Development, Inc., USA, epoxy-based SMP (TEMBO®); Lubrizol Advanced Materials, USA, aliphatic TPU-based SMP (Tecoflex®); MedShape, Inc., USA, polyether ether ketone-based SMP (Morphix®, ExoShape®); and Norland Products Inc., USA, UV-curable polyurethane-based SMP (NOA-63).

SMPs have taken paramount interest in recent times because of their potential application in numerous fields, for example, smart actuators, aerospace engineering, automotive engineering, textile engineering and most prominently as smart biomedical devices [10–17]. Ni–Ti alloy (Nitinol) is the most extensively used shape memory material because of its good shape memory performance, excellent mechanical properties, good biocompatibility and so on [18]. However, the shape memory alloys (SMAs) have some disadvantages such as low deformation rate, high stiffness, high weight, high cost and nonbiodegradability. SMPs have some advantages such as easy processing, lower density, larger recoverable strain, tunable transition temperatures, multiple shape recovery ability, lower cost and lower toxicity when compared to the SMA (Table 1.1). SMPs are smart materials that can transform their shapes in a controlled manner when exposed to a suitable external stimulus such as heat, light, solvent, pH, magnetic field and electric field [19–24]. SMPs have the ability to memorize specific macroscopic and permanent shape. They have

https://doi.org/10.1515/9783110570175-001

Table 1.1: Comparison of the properties of SMP and SMA.

Physical properties	SMP	SMA
Density (g/cm^3)	0.9–1.1	6–8
Extent of deformation (%)	250–800	<8
Force required for deformation (MPa)	1–3	50–200
Recovery temperature (°C)	25–90	−10–100
Recovery stress (MPa)	1–3	150–400
Recovery speed (s)	>1	<1
Processing conditions	<200 °C, low pressure	>1,000 °C, high pressure

the ability to fix a temporary shape under specific temperature and stress, and can revert back to their original shape on application of appropriate external stimulus. The recovery of original shape from the deliberated fixed temporary shape is known as shape memory effect (SME). SME is quantified by the shape fixity and shape recovery function. The shape memory functions, i.e., the shape fixity and shape recovery of polymer network, are the freezing and activation of the macromolecular chains below and above a transition temperature (T_{trans}), respectively. The extent to which the switching segment is able to fix the temporary shape is called the shape fixity. The extent to which it recovers the original shape from the temporary shape on exposure of stimulus is called the shape recovery. The shape recovery time is a shape memory property which refers to the time that SMP used to take recovers the permanent shape from its fixed temporary shape. The T_{trans} can be either a glass transition temperature (T_g) or a melting temperature (T_m) [25, 26]. For amorphous polymers, T_g is considered as T_{trans}, whereas for crystalline polymers, T_m is considered as T_{trans}. Moreover, for semicrystalline polymers, either T_g or T_m can be used as T_{trans}. However, T_m is being considered as the preferred T_{trans} because it is sharper than the T_g [27]. Shape memory behaviors observed through shape memory cycle consist of three successive steps. The shape memory cycle is referred to the evolution of stress (or force), strain and temperature during thermomechanical cycling of an SMP.

1. Deformation: The SMP is first deformed to a predetermined stress or strain at the deformation temperature (T_d, $T_d > T_{trans}$), at which the material becomes soften (modulus drop).
2. Fixing: The deformed sample is cooled under constant stress or strain below T_{trans}. This causes the material to attain a more rigid state, in this case the semicrystalline state, therein immobilizing the constituent polymer chains and allowing fixing or freezing of the deformation as latent strain energy.
3. Recovery: The temporarily deformed sample, i.e., fixed shape is reheated above T_{trans} without any stress (unconstrained). The stored strain energy is then released owing to the regained chain mobility and it returns to its original shape. The elastic strain generated during deformation is the driving force for the shape recovery.

1.2 Molecular mechanism of SME

Most of the polymers perform SME because polymer chains have tendency to form network structures. However, limited numbers of SMP are used in practical applications, due to their structural drawbacks (too low or too high T_{trans} or poor mechanical properties). The molecular mechanism of thermally induced SMP through which SME observes is shown in Figure 1.1. Cross-links and switching segments are the key components at molecular level to exhibit the SME. The cross-links or netpoints determine the permanent shape of SMP which inhibits slippage of the macromolecular chains under large deformations, whereas switching segment absorbs the external stress and maintains the temporary shape [6, 28]. The netpoints may be either chemical (covalent bonds) or physical (secondary interactions or entanglements) in nature. Physical cross-linking is obtained in a polymer whose morphology consists of at least two segregated domains. In such morphology, the domain related to the highest thermal transition temperature acts as physical netpoints which is called a hard segment. The domain associating with the second highest thermal transition acts as molecular switches, which is called a switching domain. The molecular switches must be able to fix the deformed shape temporarily under environmental conditions by forming additional reversible cross-links. These additional temporary cross-links can be formed by physical interactions or by chemical bonds which prevent the recoiling of the deformed chains. Physical cross-links are obtained through a solidification of switching segments by vitrification or crystallization that enables the storage of

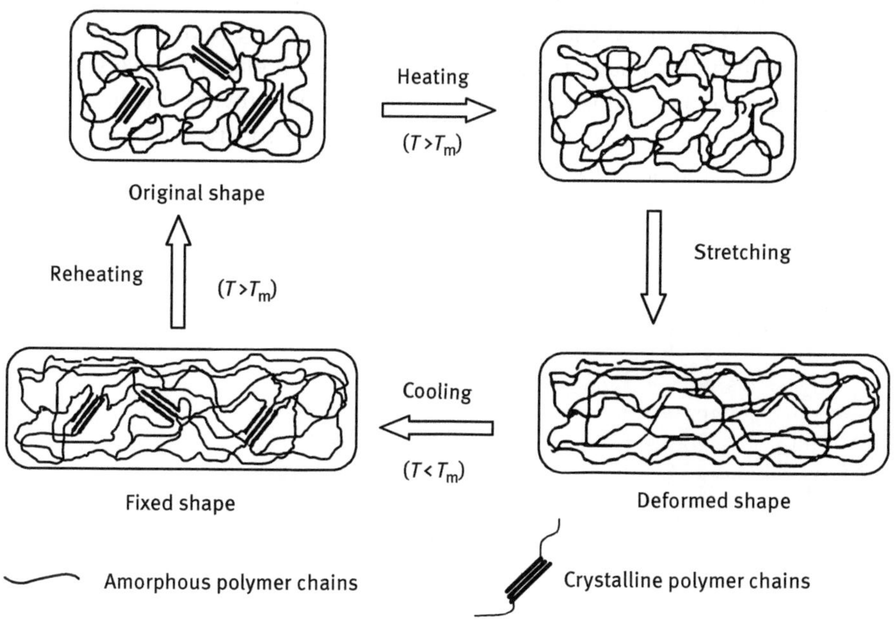

Figure 1.1: Molecular mechanism of thermally induced SMP.

entropic energy in the system [18]. After reheating, the crystallites will melt or the glassy domains will soften, resulting in the release of entropic energy by recoiling the chains that drive the material to return to its original shape. Chemical cross-links are obtained by reaction of two functional groups under formation of a chemical bond that is able to cleave by external stimulus can be used as molecular switches.

The aforementioned discussion implies that the polymer SME is an entropic phenomenon. From the molecular point of view, it is an activation and freezing of molecular chains by a temperature change. In the permanent shape, the macromolecular chains within the SMP network structure are at the lowest energy (highest entropy) state, which is thermodynamically favorable. The conformation of the chains would change and raise the energy state of the material, when macroscopic deformation takes place in the permanent shape. At the deformation stage, when it is cooling, freeze the molecular chains mobility, which led to the kinetic trap to maintain the system in the high energy state. The entropic energy is released, once the kinetic trap is removed by reactivating the chains mobility (e.g., by heating) and as a consequence recover the original shape by driving the molecular chains to return to their lowest energy (highest entropy) state. The storage and release of entropic energy is responsible for the shape fixing and recovery of the SMP. To satisfy the entropic phenomenon, the system should not have slippage of entire polymer chains that result in macroscopic deformation but no entropic change. An illustration of shape recovery of shape memory polyurethane in hot water at 60 °C is shown in Figure 1.2.

The modulus (E)–temperature (T) behavior of the materials is of basic importance with regard to understanding the shape fixity and shape recovery effect. A high glassy state modulus (E_g) of material exhibits high shape fixity during cooling and unloading, whereas high rubbery state modulus (E_r) leads to high elastic recovery at high temperature. The large modulus ratio (E_g/E_r) of the material provides the high shape recovery effect. Further, a sharp transition from glassy to rubbery state leads to the material sensitive to temperature variation.

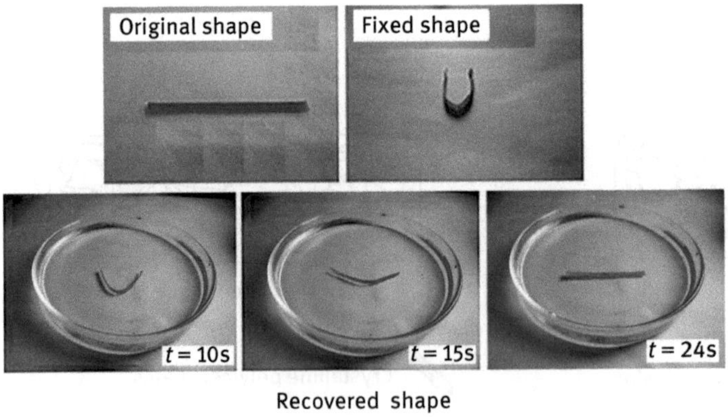

Figure 1.2: An illustration of shape recovery of shape memory polyurethane in hot water at 60 °C.

1.3 Hyperbranched SMPs

Recently, hyperbranched SMPs have taken wonderful interest because of their unique characteristics [29–32]. Hyperbranched SMPs show better shape memory properties than their linear analogue. Hyperbranched polymers are highly branched macromolecules having significantly less regular structural architecture, number of surface defects and missing branches. Hyperbranched polymer was first introduced by Kim and Webster in 1988 when the authors synthesized soluble hyperbranched polyphenylene with AB_x-type monomers [33]. Hyperbranched polymers show some unique properties such as high solubility and chemical reactivity, low melt and solution viscosity, good compatibility, compact three-dimensional nonentangled globular structures and good shape memory behaviors [32, 34]. The plenty of functional groups present in the hyperbranched polymers render the opportunity for further modification, and the type of end groups governs the extent of properties. Shape memory properties of hyperbranched polymers can be tuned as a function of degree of branching (DB), core structure and type of functional end groups. Furthermore, hyperbranched polymers exhibit unique polymeric attributes such as a wide molar mass distribution, isomerism and an irregular growth with a statistical distribution of the functional groups in the structure. The most important attribute of hyperbranched polymers is their DB, which can be calculated as $DB = D + T/D + L + T$, where D, T and L are the dendritic, terminal and linear units in the polymer structure. DB is 1 for dendrimers and <1 (0.5 for a statistical growth) for hyperbranched structures. Usually, two synthetic methods are being used for the synthesis of such hyperbranched polymers. These are single-monomer methodology (SMM) and double-monomer methodology (DMM). Among different approaches of SMM, step-growth polycondensation method of ABn monomers is extensively used to synthesize a wide range of hyperbranched polymers. Again, $A_2 + B_3$ approach of DMM is preferred to use due to its several benefits. These include easy availability of the reactants, easy preparation and tuning of various properties including shape memory by using long segmented reactant. Thus the use of $A_2 + B_3$ approach may render a unique route and the resulted polymer exhibits unique architectural features with unique shape memory properties. However, this approach has one major disadvantage of gelation. This can be avoided by performing the reaction under low concentration of monomers, slow addition of monomers, adjusting the reaction time and temperature and/or stopping the reaction before the critical point of gelation [33, 35].

The shape memory behaviors of copolymeric elastomers derived from lignin and glycerol-adipic acid-based hyperbranched prepolymers have been studied by Li et al. [36]. They prepared a series of hyperbranched prepolymers, wherein the first step was melt condensation of glycerol (Gly, $B^2B^3{}_2$) and adipic acid (AA, A_2), and the prepolymer structure was also modified by reacting with commercially available diisopropanolamine (DIPA, $DB^4{}_2$) and tris(hydroxymethyl) aminomethane (THAM, $CB_3{}^1$) units. In the second step, the hyperbranched prepolymers were reacted with lignin to prepare thermoresponsive shape memory copolymeric elastomers in the same pot through melt polycondensation. The materials exhibited good shape fixity and shape recovery behavior. The glycerol-based polyester and polyester amide acted as soft segment and the lignin acted as netpoint segment. It was found that T_{trans} was affected by the amount of carboxylic acid as well as the addition of DIPA or THAM to the glycerol-based prepolymer formulations. The higher T_{trans} was observed with the higher content of carboxylic acid, DIPA or THAM in the prepolymer.

SME of epoxy thermosets prepared by cross-linking of diglycidyl ether of bisphenol A with a mixture of commercially available hyperbranched poly(ethyleneimine) and polyetheramine was reported [37]. The hyperbranched polymer exhibited higher storage modulus. The storage modulus at the rubbery state increased with the increase of the content of poly(ethyleneimine). This is expected to the increase of the cross-linking density in the presence of poly(ethyleneimine). The materials showed significant differences of storage modulus at the glassy and rubbery regions, which is crucial for the SME. The materials display excellent shape memory behaviors with maximum shape recovery and shape fixity of about 98% and shape recovery rate of 22%/min. The results demonstrated that hyperbranched poly(ethyleneimine) served as a cross-linking agent is used to improve mechanical and shape memory properties with different effects depending on the cross-linking density. It was suggested that the local deformation of the internal structure of hyperbranched poly(ethyleneimine) in the network, with short ethylene segments within, has a lower contribution to conformational, entropic deformation and can therefore relax immediately after stress release. The shape recovery and shape fixity ratios found to decline slightly with increasing hyperbranched poly(ethyleneimine) content because it restricted chain conformational changes during deformation and molecular dynamics. It was concluded that the shape memory properties of epoxy can be improved through modifying with controlled amount of hyperbranched polymer because of its effect on network homogeneity.

SME of poly(ε-caprolactone) (PCL)-based hyperbranched shape memory polyurethanes synthesized via $A_2 + B_3$ has been investigated [38]. The storage modulus ratios (E' ratios) of hyperbranched polymers were found to be significantly high when compared with the linear analogue. The hyperbranched polymer exhibited higher than δ value which corresponds to strain energy dissipated by viscous friction. As a consequence, it can be concluded that the hyperbranched shape memory polyurethanes are more likely viscous than elastic and will possess significant damping property. The hyperbranched shape memory polyurethanes exhibited 100% shape recovery when compared with the linear counterpart as a result of increased stored strain energy of hyperbranched polymers. Bayan and Karak [39] reported the multistimuli shape memory behaviors of aliphatic hyperbranched polyurethanes synthesized from castor oil modified polyol containing fatty amide triol, glycerol, diethanolamine and monoglyceride of sunflower oil through $A_x + B_y$ ($x, y \geq 2$) approach. The formulated aliphatic hyperbranched polyurethanes exhibited excellent multistimuli actuated shape recovery behavior (up to 97%) under direct sunlight (10^5 lux), thermal energy (50 °C) and microwave irradiation (450 W). The hyperbranched polyurethanes showed faster shape recovery as compared to the linear counterpart. They also reported that the hyperbranched polyurethanes possess faster recovery under the actuation of thermal and microwave irradiation as compared to the actuation of sunlight.

SME of hyperbranched polyurethane has been reported to be synthesized from *Mesua ferrea* L. seed oil as a biobased chain extender, triethanolamine as a core moiety (at different weight percentages) and toluene diisocyanate (TDI) by a pre-polymerization method through $A_2 + B_3$ approach [40]. The reaction mechanism of this hyperbranched polyurethane is shown in Figure 1.3. Nuclear magnetic resonance study confirmed the formation of hyperbranched structure with DB 0.9. The hyperbranched polyurethane exhibited higher shape recovery ratio and faster shape recovery as compared to the linear analogue. The shape recovery ratio found to increase with the increase of the amount of core moiety of hyperbranched polymers. This excellent shape recovery of the hyperbranched polymers may be attributed to the increased stored strain energy of system due to homogenous distribution of hard segments and core moiety, triethanolamine in the structure and the enhanced secondary interaction in the polyurethanes. Similarly, Duarah et al. [41] also studied the effect of core moiety on the SME of hyperbranched polyurethane synthesized from starch-based polyol as core moiety, PCL as soft segment, butanediol as chain extender and TDI. They synthesized the starch-based polyol by reacting starch with bisphenol-A and epichlorohydrin. It was observed that hyperbranched polyurethane showed better shape recovery behaviors as compared to the linear analogue near body temperature (37 ± 1 °C). Similarly, the shape recovery is observed to increase with the increase in the weight percentage of branch generating moiety. The better shape recovery of the hyperbranched structure can be attributed to the enhanced secondary interactions and the physical or virtual cross-links among the new orientated polymer segments.

1.4 Interpenetrating SMPs

Interpenetrating polymer networks (IPNs) have been employed to develop new SMP with improved properties [42, 43]. IPNs can be described as a special kind of the polymers, fabricated by a combination of two or more polymers via intermolecular permanent entanglements in which one is synthesized, physically cross-linked or polymerized in the immediate presence of other [44, 45]. The entanglements cannot be separated unless the chemical bonds are broken. Based on the synthesis and arrangement pattern, IPN can be classified as sequential, simultaneous, semi-IPN and so on [46, 47]. In sequential IPN, after the completion of polymerization of the first polymeric component network, the second component network is polymerized. On the other hand, simultaneous IPN is prepared by a process in which both component networks are polymerized at the same time. In semi-IPN, one polymeric component of the assembly

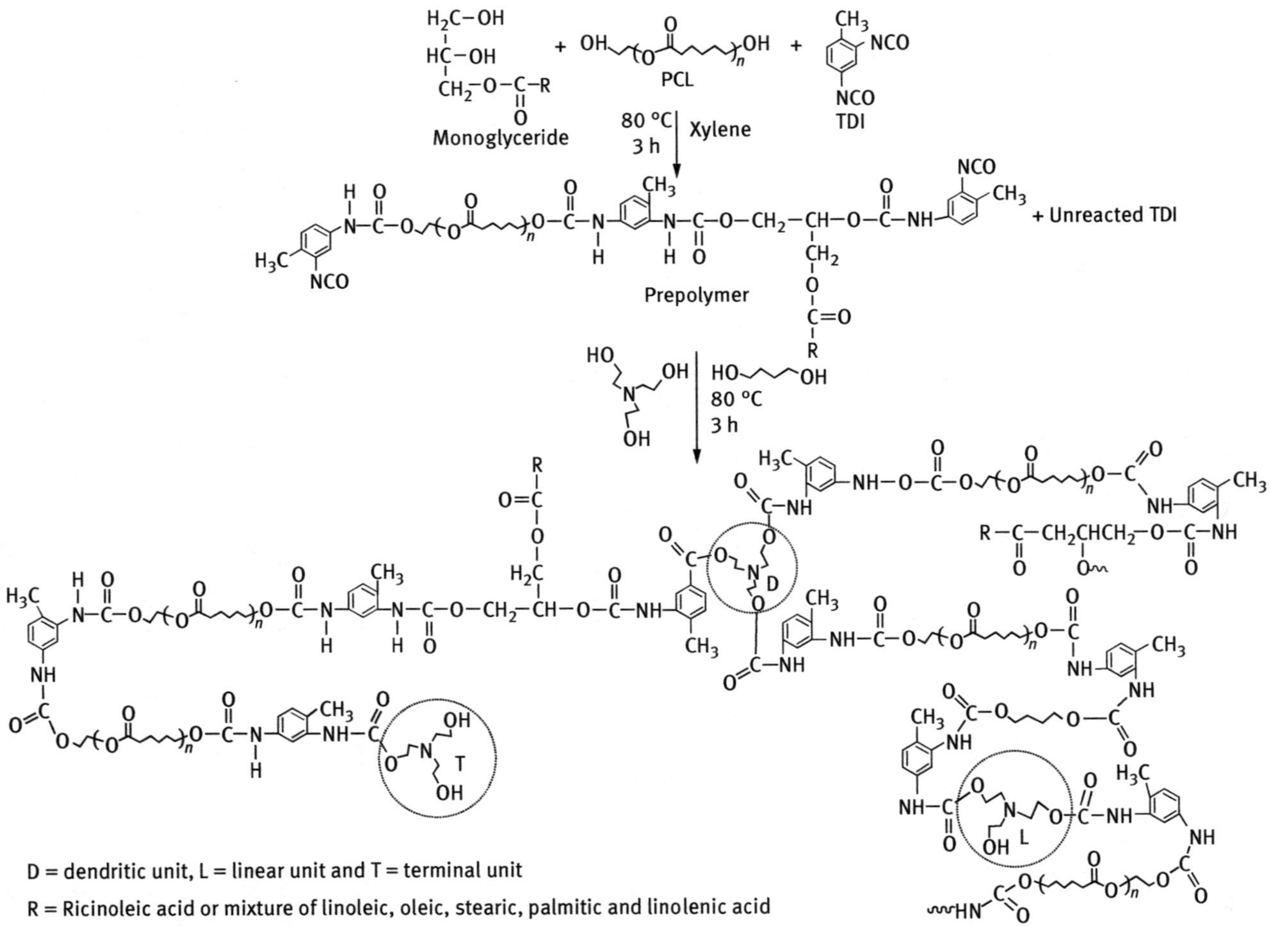

D = dendritic unit, L = linear unit and T = terminal unit

R = Ricinoleic acid or mixture of linoleic, oleic, stearic, palmitic and linolenic acid

Figure 1.3: Reaction mechanism of shape memory hyperbranched polyurethane synthesized through $A_2 + B_3$ approach.

is cross-linked, whereas the other in a form of linear structure. In the interpenetrating SMP, the produced netpoints of both polymer networks determine the permanent shape, whereas polymer networks serve as the switching domains. The shape memory properties of IPN depend on the weight ratio of the IPN components, network structure, cross-linker concentration and morphology of the resulting network structure.

Shape memory semi-IPN using crystalline poly(ethylene oxide) (PEO) and cross-linked poly(methyl methacrylate) (PMMA) have been developed [48]. The effect of cross-linker concentration and recovery temperature on the shape recovery rate of semi-IPN has been studied. The recovery time was found to increase with increasing cross-linker concentration for all the recovery temperatures, whereas the shape fixity tended to decrease with decreasing cross-link density. It was observed that semi-IPN sample having a composition of PMMA/PEO = 68/32 and cross-linker concentration of 4 wt% exhibited the best shape memory properties. It was suggested that IPN should have optimum compositional ratio and cross-linker concentration to exhibit excellent SME. A novel poly(ester urethane)/poly(ethylene glycol) dimethacrylate (PEGDMA) IPN has been developed [49]. The IPN exhibited only one T_g value and it was decreased with the increase in the content of PEGDMA. The IPN showed good shape recovery and shape fixity ratio of more than 93% and the T_{trans} can be adjusted in the range of −23 to 62.8 °C. The chemical cross-linking points served as a fixed phase to memorize the original shape, while the amorphous domains of miscible PLGA and PEG acted as a reversible phase. The switch temperature of the IPN could be adjusted to around body temperature which is potential for clinical applications. Liu et al. have developed PMMA-PEG shape memory semi-IPN [42]. The IPNs were synthesized by radical polymerization and cross-linking of PMMA with EGDMA in the presence of linear PEG. The IPN exhibited two phases, namely the crystalline PEG and the amorphous PMMA-PEG complex. The IPN exhibited two independent SMEs at the T_m of the PEG crystals and the T_g of the semi-IPN. For the SME at the T_m (T_{trans}) of PEG crystals, the PMMA network serves as the fixing phase, while PEG crystal serves as the reversible phase. However, for the SME at the T_g (T_{trans}) of the semi-IPN, the chemical cross-linked points act as the fixing phase, whereas the PMMA-PEG complex phase acts as the reversible phase. The IPN showed excellent shape recovery behaviors of more than 90%, and the large difference in modulus above and below T_g is the key for the good shape memory behavior. The author also designed novel shape memory PMMA-co-*N*-vinyl-2-pyrrolidone (PVP)/PEG semi-IPN based on hydrogen bonding [50]. The IPN exhibited shape recovery ratio of 99%. The shape recovery rate found to increase with the increase in molecular weight of PEG. In the IPN, the fixing phase was the chemical cross-linked points, whereas the reversible phase was the PEG-PVP complex phase. It was concluded that SME can be easily tuned by varying the molecular weight of the PEG and the cross-linking density.

Multiple SMEs from dual to quintuple were observed in PMMA/PEG semi-IPN containing broadened glass transition and crystalline segments. The PMMA/PEG semi-IPNs were synthesized by in situ homopolymerization technique and cross-linking of MMA in the presence of linear PEG [51]. The PMMA/PEG IPN showed quintuple SME by utilizing three separated temperatures (110, 70 and 50 °C) within the glass transition and additional T_m of PEG (35 °C) as fourth T_{trans}. These demonstrated that the cause of multi-SME of the semi-IPN, PMMA network and linear PEG was forcedly interlocked, making the molecular movements constrained, as a consequence in the relative hysteresis of polymer molecular relaxation. As a result, the T_g of semi-IPN would span a broader temperature range that rendered an opportunity to generate the multigradient T_g (i.e., T_g-high and T_g-low). Again, the presence of cross-linking in networks contributed to the formation of aggregate PEG phase having a T_m. All polymer chain segments in semi-IPN were flexible at T_g-high. The first temporary shape was achieved by cooling from T_g-high to T_g-low, which led to the verification of polymer chain segments in amorphous PMMA/PEG mixed phase. The second temporary shape was achieved by subsequent cooling from T_g-low to T_m, the entire amorphous PMMA/PEG phase became completely glassy. At T_m, the chain segments in aggregate PEG phase were still flexible. The third temporary shape was fixed through physical cross-links by cooling to the crystallization temperature (T_c). Once the material was heated in a stepwise fashion to T_g-high again, the energy frozen previously was activated portion by portion, which enabled the three-step shape recovery to restore the permanent shape. They demonstrated that by employing interpenetration PMMA/PEG, multi-SME can be achieved by simply choosing the suitable temperatures from the T_g and additional T_m.

1.5 Blend SMPs

Polymer blends have gained tremendous interest because of their improved mechanical, thermal, electrical, biodegradability and shape memory properties [52–58]. By blending two or more polymers one can fabricate new SMP [59]. Further, the SME of SMP can be tuned by blending it with other suitable polymer. Phase morphology, structure as well as amount of each phases, and crystallization condition have strong influence on SME of SMP. The switching temperature of SMP can be adjusted on demand by blending. The shape memory properties, namely shape recovery and shape fixity, have strong influence on the phase separated structure of the blends that are related to the entangle structure, presence of cross-links, hydrogen or ionic bonds or physical intermolecular interactions between the polymers. The blending of miscible polymers and immiscible polymers, by both the methods is attractive to develop new SMP.

1.5.1 Miscible blend SMPs

Blending of miscible polymers is an adequate technique to develop novel SMP because fabrication is easy and widely applicable. The blending can be accomplished with different approaches such as SMP/polymer, amorphous polymer/crystalline polymer and crystalline polymer/crystalline polymer.

1.5.1.1 SMP/polymer blend

The blending of an SMP with its miscible polymer, the mechanical properties and the switching temperature of the SMP can be adjusted as desired. Among different SMPs, segmented polyurethanes are well-known SMPs and have gained utmost interest due to its excellent shape memory properties. Shape memory polyurethanes blended with other miscible polymers have been reported. Segmented TPUs synthesized from diol-terminated PCL, hexamethylene diisocyanate and 4,4′-dihydroxy biphenyl blended with poly(vinyl chloride) (PVC) have been reported [60]. PVC is miscible with soft segment PCL in TPU and the switching temperature of the blends, i.e., T_g of miscible amorphous domain of PVC/PCL could be tuned by varying the composition. Jeong et al. [61] also reported the SME of segmented TPU blended with phenoxy resin. The miscible domain of PCL/phenoxy resin acts as reversible phase and the hard domain of the TPU acts as fixed structure memorizing the original shape. The switching temperature could be tuned, which lied between the T_g of PCL and phenoxy resin. The hysteresis of the shape recovery was found to diminish when the content of hard segment in blend was increased and the block length of the TPU was decreased. SME of TPU synthesized from 4,4′-methylenebis(phenyl isocyanate), poly(tetramethylene)glycol and 1,4-butanediol blended with polybenzoxazine (PB) have been reported [62]. The benzoxazine produced miscible blends with polyurethane prepolymer and some degree of reactions between phenolic –OH groups in PB, and isocyanate groups in chain extended polyurethane took place during thermal curing of benzoxazine. The TPU/PB blend exhibited excellent shape memory behaviors by means of higher shape recovery force, faster shape recovery and greater shape recovery ratio than polyurethane.

1.5.1.2 Amorphous polymer/crystalline polymer blend

The blending of amorphous and crystalline polymer is a new approach to develop SMP. The crystalline polymer acts as the cross-linking structure, and the amorphous polymer serves as the switching polymer. Mather [63] patented the miscible amorphous/semicrystalline blend SMP. The amorphous polymers include poly(methyl acrylate), poly(ethyl acrylate), atactic PMMA and poly(vinyl acetate), and the crystalline polymers include poly(vinylidene fluoride) (PVDF), polylactide, poly(hydroxybutyrate), poly(ethylene glycol) (PEG), polyethylene, PVC and poly(ethylene-co-vinyl acetate). The polymer blends of both the components were entirely miscible at the molecular level, and only single T_g was observed for all blend ratios. The rubbery modulus can be tuned by adjusting the degree of crystallinity of the blends with varying the blend compositions, which is important for SME. Crystalline/amorphous blend SMP of PVDF/acrylic copolymer (ACP) has been developed [64]. The blends of all the compositions showed only one T_g, confirming molecular miscibility between the PVDF and ACP. The miscibility is due to the interactions of carboxyl groups of ACP with the CF_2 groups of PVDF. The PVDF/ACP blends with 40–50 wt% of PVDF exhibited excellent shape memory functions with a switching temperature of approximately 45 °C. It was proposed that the small crystals of PVDF in the blends acted as netpoints to fix the shape and tie molecules, i.e., amorphous chain connects these netpoints to form a nice elastic network that served as switching segments for the shape recovery. The tie molecules resist molecular slippage upon deformation, and the small PVDF crystals serve as the fixed phase for the shape memory performance. For the blends with lesser wt% of PVDF, the PVDF crystals are isolated from each other and are embedded in the amorphous matrix, resulting in significant permanent deformation owing to molecular slippage during mechanical stretching. For the blends with higher wt% of PVDF, PVDF usually crystallizes into large spherulites; this crystallization breaks the molecular structure and causes the permanent deformation during mechanical stretching. Thus, these samples exhibit relatively low SME. The optimum amount of crystals and tie molecules is critically important to show the excellent SME.

1.5.1.3 Crystalline polymer/crystalline polymer blend

The blending of two crystalline polymers is a useful method to develop blend SMP. One crystalline polymer acts as a fixing phase and the second crystalline polymer acts as a reversible phase. Blend SMP of two crystalline polymers poly(*p*-dioxanone) (PPDO) and PCL with poly(alkylene adipate) mediator as a compatibilizer has been developed [65]. The crystalline PPDO acting as a hard segment determines the permanent shape and the PCL acting as a soft segment determines the temporary shape. The blend SMP exhibited excellent shape memory properties, and the mechanical properties can be varied systematically. The switching temperature of the blends, i.e., melting point associated with the PCL, is almost independent of the weight ratio of the two blend components. In this approach, the elastic properties of the blend SMP can be tuned as necessitated without affecting the switching temperature. A novel blend SMP from two crystalline polymers, high-density polyethylene (HDPE) and poly(ethylene terephthalate) (PET) with ethylene-butyl acrylate-glycidyl methacrylate terpolymer (EBAGMA) as a reactive compatibilizer, has been developed [66]. The HDPE phase with the soft molecular chains acts as the reversible phase and the PET phase with relatively rigid molecular chains serves as the fixing phase. EBAGMA compatibilizer effectively improves the compatibility between the HDPE and PET due to the structural similarities among them, and the epoxy groups of the glycidyl methacrylate of EBAGMA may react with the hydroxyl and carboxyl groups of PET. The blend showed excellent SME with 5 phr EBAGMA and the ratio of HDPE/PET with 90/10. A blend SMP based on ethylene octene copolymer (EOC) and ethylene propylene diene terpolymer (EPDM) rubber has been reported [67]. EOC acts as hard domain and thus provides stiffness and reinforcement, whereas EPDM acts as soft segment and is responsible for the thermoelastic behavior of the blend. The blend having 75% EPDM and 25% EOC showed the highest percentage of shape recovery, whereas the blend containing 25% EPDM and 75% EOC showed the lowest shape recovery. The blend with higher amount of EPDM results in higher shape fixity because of the formation of higher level of EPDM crystallites and the melting of all the crystallites in the presence of heat and also the consequence of highest percentage of shape recovery. The better shape memory behavior of the EOC-EPDM blends even without any cross-linking is primarily due to the high degree of compatibility between the two pristine polymers because of the structural similarities and in turn facilitate the formation of enough physical entanglements between the two phases.

1.5.2 Immiscible blend SMPs

When the blending polymers are not mixed at the molecular level is defined as the immiscible blend. Immiscible blends exhibit a phase separated structure, which influences the SME of blends. Immiscible blend of one elastomer and the other crystalline polymer acting as switching polymer display good shape memory properties. A novel blend SMP of styrene-butadiene-styrene (SBS) triblock copolymer and PCL have been reported [68]. SBS acts as an elastomer, whereas semicrystalline PCL acts as switching polymer. The shape recovery ratio dropped from 100% to zero when the PCL content increased to 50 wt%, which is due to the less amount of elastic phase. The shape fixity ratio increased from 40% to 100% after PCL content 30 wt%. It signified that the networks of SBS provide the shape recovery performance while the crystallization of PCL contributed to shape fixing performance. It was suggested that to develop this type of blend SMP, two types of immiscible components are required. One is the elastomer, which can be any rubberlike or thermoplastic elastomers whereas the other is the switching polymer, which can be any crystalline polymers. Through suitable design of the immiscible phase morphology, an ideal SMP can be achieved with both good stability and performances. Weiss et al. [69] have designed blend SMP from an elastomeric ionomer with low molar mass fatty acids (FA) and their salts. Nanophase separation of the ionomer was used to build up a cross-linking network, and FAs (salts) were used to make a secondary network. The switching temperature can be easily tuned by choosing FAs with different melting points. Ionomer possesses a strong intermolecular bond between the FA crystal and the polymer that will behave as a physical cross-link below switching temperature and allow reshaping of the material above switching temperature. The blend showed poor SME in the absence of ionomer because interactions between polymers and FA crystals are not strong enough to afford cross-link.

1.6 Biobased SMPs

Depletion of petroleum resources, environmental legislation and waste disposal problems motivates the utilization of bioresources such as vegetable oils, starch, cellulose, latex, sugar and so on as raw materials for the synthesis of SMP [70–72]. This is due to the fact that bioresources have several advantages such as easy availability, relatively low cost, versatility in structure and property, inherent biodegradability and environmentally benign. Biobased SMPs gain special attention for biomedical applications because of their structural characteristics, biocompatibility and expected biodegradability. The shape memory properties of biobased SMP can be tuned by varying the concentration and opt the chemical composition of the biobased component. The elasticity, modulus, cross-linking density and switching temperature of biobased SMP can be adjusted by suitable design of the biobased component. The SMPs derived from biomasses or their extracts are of immense importance and extremely required [52, 73–79]. However, direct radical or cationic polymerization of the majority of plant oils is structurally difficult due to their nonconjugated and internal double bonds, which result in the formation of low-molecular-weight viscous liquid polymers. Some of the vegetable oils inherently contain functional groups (castor oil/hydroxyl group; vernolin oil/epoxy group); these can be directly used for the synthesis of numerous SMPs. The natural oils, and mostly the unsaturated ones, are of special interest because a variety of reactions could be performed from their different functional sites in order to obtain required biobased monomers. Various reactions through double bond reactions such as hydroformylation, epoxidation, metathesis and so on, and through ester bond reactions such as transesterification and transamidation are performed to functionalize the oils (Figure 1.4) [80–82]. Hydroxylation of vegetable oils is one of the good approaches to get the required polyols to be used as diol chain extenders in the synthesis of shape memory polyurethanes. Properties of the vegetable oil-based SMP depend on attributes such as physical and chemical structures, which include the degree of unsaturation, number of hydroxyl groups present in the polyol, length of the FA chains and position of hydroxyl groups in the FA chain. The structures and compositions of different vegetable oils are presented in Table 1.2. Vegetable oils have taken important role to develop shape memory epoxy where oil acts as reactive diluents as well as a soft segment.

Biobased SMPs synthesized from soybean oil, styrene and divinylbenzene by the cationic polymerization reaction have been reported by Li and Larock [83]. The reaction mechanism of this SMP is shown in Figure 1.5. With the increase of soybean oil concentration, deformability is found to increase while shape fixity behavior is found to decrease. The polymer behaves like an elastomer, when the soybean oil concentration exceeds 50%. The increased soybean oil concentration lowers the degree of cross-linking, which leads to a greater number of conformations that the polymer can adopt upon being loaded and deformed, as a result improving the deformability as an elastomer above T_g. The increase in soybean oil concentration reduced the T_g as a result of decreased rigidity of the polymer chains and cross-link density. Therefore, the micro-Brownian movement of the polymer chains cannot be effectively frozen, as a consequence reduce the percentage of shape fixity. The polymers exhibited 100% shape recovery, signifying that the cross-link density was high enough to effectively store and release the elastic strain energy during the shape memory process. The

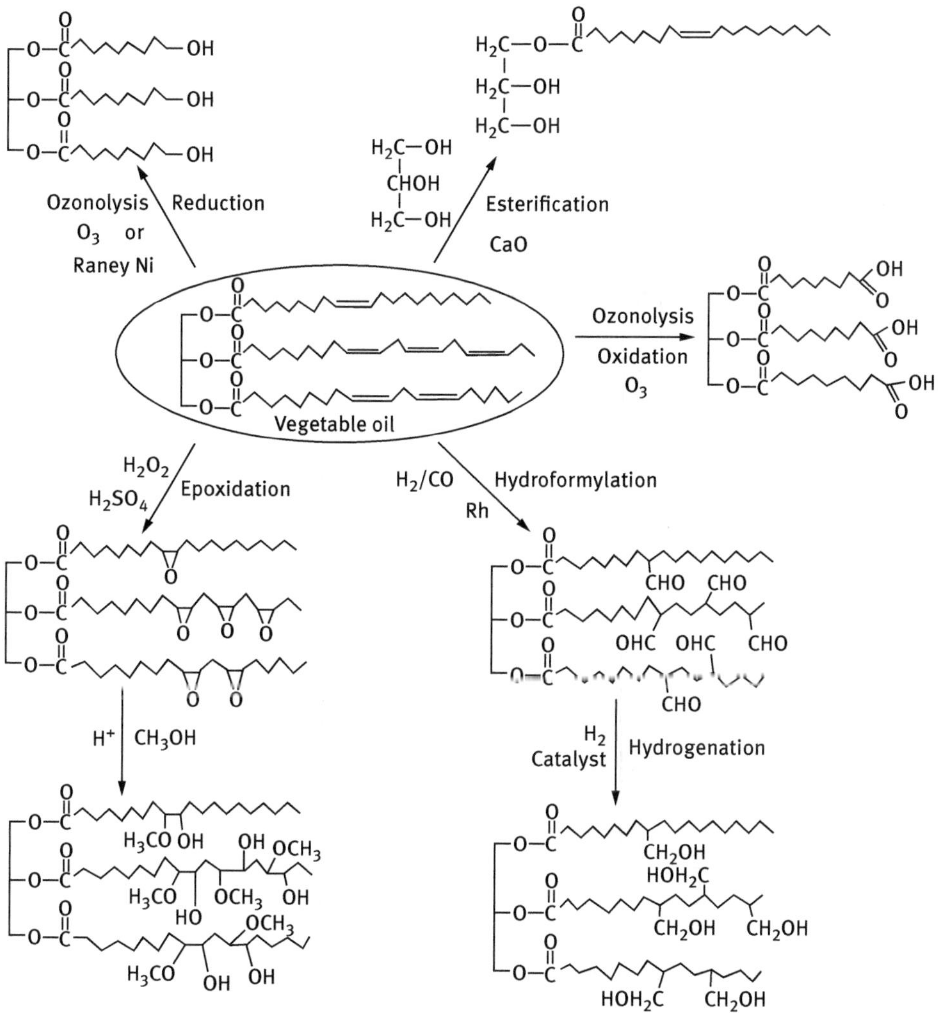

Figure 1.4: Various reactions for functionalization of vegetable oil.

effect of different types of soybean oils such as regular soybean oil, low saturation soybean oil and conjugated low saturation soybean oil on the shape memory properties has been investigated. The conjugated low saturation soybean oil has higher reactivity toward the cationic polymerization reaction, as a result the polymer possesses a higher degree of cross-linking and T_g than the other soybean oil polymers. Thus the conjugated low saturation soybean oil polymer showed higher shape fixity behavior and lower deformability. Good SMP should have suitable combination of cross-link densities and T_g values, which are moderate cross-link densities and high T_g values well above ambient temperature. It was concluded that by attaining appropriate combinations of glass transition temperatures and cross-link densities through structural

Table 1.2: Structures and composition of different vegetable oils

Fatty acid	Structure	*Mesua ferrea* L. seed	Sunflower	Castor	Palm	Soybean	Linseed
Palmitic acid (%)	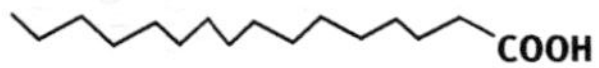	15.9	5.2	–	45	11	6
Stearic acid (%)	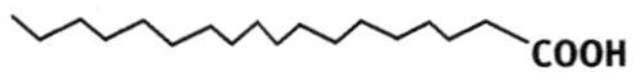	9.5	2.7	–	4	4	3
Oleic acid (%)	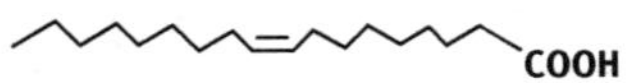	52.3	37.2	–	39	23	17
Linoleic acid (%)	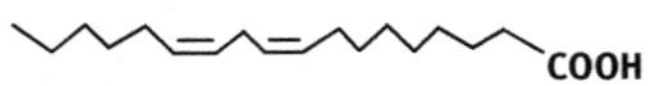	22.3	53.8	–	11	53	14
Linolenic acid (%)	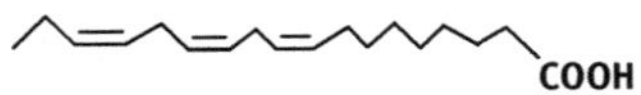	–	1.0	–	1.0	8	60
Ricinoleic acid (%)	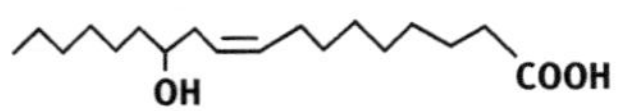	–	–	95	–	–	–

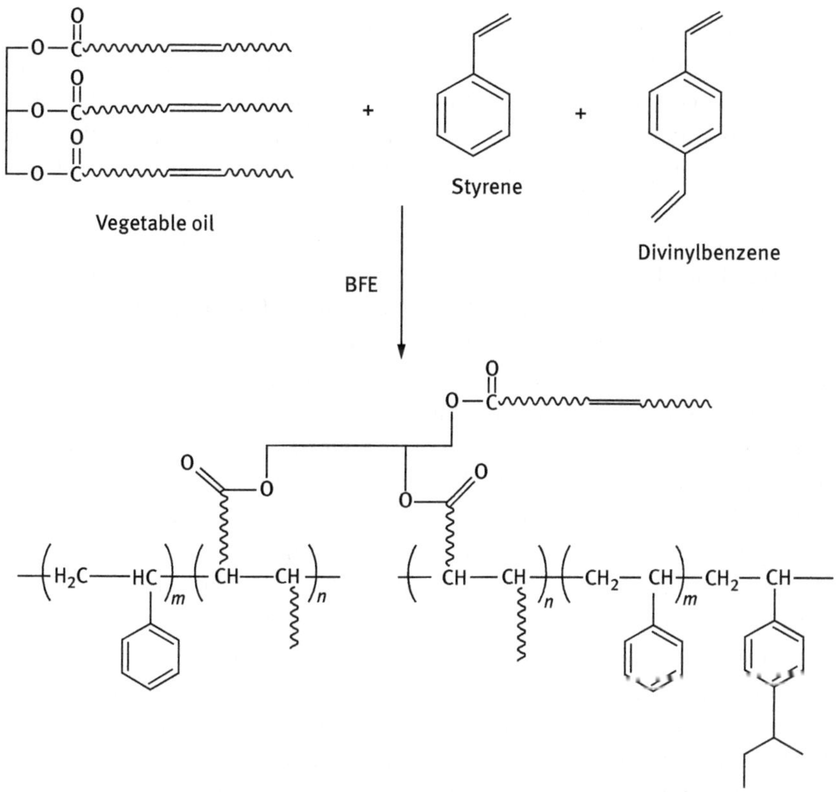

Figure 1.5: Biobased SMP synthesized from vegetable oil, styrene and divinylbenzene.

design of the polymer, soybean oil-based SMP can be prepared with high deformability, degree of shape fixity and final shape recovery. Plant oil-based SMP have also been synthesized from the epoxidized soybean oil (ESO) and PCL [77]. Poly-ESO/PCL SMPs were synthesized by an acid-catalyzed curing in the presence of PCL and to form a semi-IPN. The materials exhibited good shape recovery behavior of more than 90%. The crystal of PCL fixes the temporary shape of the sample while the elastic force of the ESO-based network polymer is the driving force of shape recovery process. The shape fixity was found to decrease with the increase of ESO content due to the decrease in the crystallinity of PCL component and the increase in internal stress of the ESO-based polymer network.

Biobased poly(propylene sebacate) SMPs from 1,3-propanediol, sebacic acid, itaconic acid and diethylene glycol have been developed [84]. 1,3-propanediol and itaconic acid could be industrially produced via fermentation of glucose, while sebacic acid could be derived from castor oil. The developed biobased SMP exhibited excellent shape fixity and shape recovery to almost 100% as a result of high elasticity ratio (glass state modulus/rubbery modulus) of the polymers. The switching temperature (12–54 °C) and recovery speed can be tuned by controlling the composition of polyesters and their curing extent. The recovery speed of SMP depends on the rubbery modulus, i.e., higher rubbery modulus generates higher elastic stress leading to higher recovery speed. It was concluded that biocompatible and biodegradable biobased SMP with suitable mechanical strength and SME will be a potential candidate for fabrication of biomedical devices. In addition, biobased cross-linked polyester SMP synthesized from isosorbide, itaconic acid and succinic acid exhibited excellent SME [85]. Isosorbide and itaconic acid are obtained from biomass and from certain microorganisms through enzymatic decomposition. Isosorbide is synthesized through a double hydrogenation of glucose. The synthesis of biobased poly(glycerol sebacate) SMP has been reported [86]. The three-dimensional cross-linked networks of the poly(glycerol sebacate) act as fixed phase while the amorphous phase of the poly(glycerol sebacate) acts as reversible phase. The polymer network showed T_g at –37 °C and broad T_m ranging from –20 to 40 °C. The material exhibited excellent SME, having shape recovery ratio above 95% within 20 s at T_{trans} 18 °C. It was suggested that this material can be employed as potential biomaterials for employing in medicine and in other applications interfacing with biological system.

Biobased stearoyl moiety modified by cellulose stearoyl ester moisture-responsive SMP have been reported [87]. The effect of different degree of substitution of cellulose stearoyl esters on morphology and SME has been studied. The materials with lower content of stearoyl groups did not show significant crystalline characteristic. However, the extent of crystallinity increased with the increase in degree of substitution of cellulose due to the stearoyl groups. Highly substituted cellulose long-chain esters with aliphatic chain lengths of more than 12 °C are able to form ordered regions through side chains. The materials with lower degree of substitution showed higher moisture-responsive shape memory behaviors. The films (degree of substitution ≈0.3) bend when they are exposed to water vapor under ambient conditions of 35 ± 2% relative humidity and 22 ± 3 °C. The films strip start to bend and fold up, after contacting with water vapor within 1–2 s. After attaining the maximal bending extent, the strip bent down within 1–2 s. This folding/unfolding phenomenon repeat rhythmically occurs with the same frequency once it is in contact with water vapor. The film surface facing water vapor absorbs water molecules, when the film is exposed to water vapor. The film expands vertically and horizontally, which leads to a vertical and a horizontal swelling force. As a consequence, a net folding force, the swelling force, is generated, which causes the film to bend. After folding up, the swollen surface of the film is in lower relative humidity environment, i.e., lower moisture content and thus water molecules quickly evaporate resulting in the release of the bending force. The film could be reversibly fold up and fall down through absorbing and desorbing mechanism of water. It was suggested that the materials can be applied for the fabrication of moisture-responsive sensors or generators for piezoelectricity.

Mesua ferrea L. oil-based shape memory polyurethanes have been demonstrated [88]. They studied the effect of different amount of monoglyceride prepared from transesterification of *Mesua ferrea* L. on the shape memory properties of polyurethanes. It was observed that shape fixity increased with the increase in the amount of monoglyceride content. They also reported the effect of different vegetable oils such as castor oil, sunflower oil and *Mesua ferrea* L. on the SME of biobased polyurethanes. They reported that castor oil-based polyurethane showed the highest shape recovery behavior as compared to others. This is due to the presence of more physical cross-linking due to the more urethane linkages in castor oil-based polyurethane, which caused storage of more strain energy in the system.

1.7 SMP nanocomposite

The demands of advanced applications require high-performance SMP materials. SMPs have several advantages over the SMAs as already discussed in the earlier section and still have some downsides, which impose barrier for broad applications. Some downsides of SMP are the low recovery stress due to the low rubbery moduli, low recovery speed due to poor thermal conductivities and low elastic modulus. Recently, SMP nanocomposites have taken tremendous interest because of the significant improvement of many desired properties like mechanical strength, thermal, electrical, recovery stress, recovery speed and other functional properties of SMP [59, 89–92]. The SMP nanocomposites dramatically improve the elastic modulus and recovery stress of SMP. Polymer nanocomposites are combination of two or more phases wherein domain of one phase lies in nanorange (1–100 nm). This represents a better alternative to the conventional filled polymer composite systems due to the large surface area to volume ratios of nanofillers, resulting in high interfacial interactions. The properties of SMP nanocomposites depend on several factors such as nature of the nanomaterial, surface functionality, aspect ratio, distribution of nanomaterial and interaction of nanomaterial in the polymer matrix. Different actuation techniques can be carried out by the incorporation of appropriate nanomaterials in the SMP [93, 94]. By the incorporation of conducting and magnetic nanomaterials, SMP can be actuated by the electric field and magnetic field, respectively. Further inclusion of radiation absorbing nanomaterials, SMP can be triggered by the radiation. Thus, SMP nanocomposites offer new opportunities to actuate the SMP via noncontact way.

Based on dimension of nanomaterials, nanocomposites can be divided into three different categories such as zero dimensional, one dimensional and two dimensional. In the first category of polymer nanocomposites, all three dimensions of the nanomaterials are in the nanoregion. The nanomaterials such as spherical nanoparticles, quantum dots and nanodots containing polymer nanocomposites fall in this category. In the second class of polymer nanocomposites, two dimensions of the nanomaterials are in the nanorange, whereas the third dimension may be a few micrometers. The examples of this class are carbon nanotube (CNT), carbon nanofiber (CNF) and nanomaterial-based polymer nanocomposites. For nanocomposites, only one dimension of the nanomaterials is in the nanorange, while the other two dimensions are in micrometer size and are categorized as two-dimensional nanocomposites. In most of the cases, the nanomaterials are in layer structures such as silicate layers and clay and such nanomaterial-based polymer nanocomposites are included in this category. Different techniques have been used to prepare the SMP nanocomposites and are discussed below.

1.7.1 Techniques for preparation of SMP nanocomposites

1.7.1.1 Solution method

In this method, nanomaterials are swelled and dispersed by solvent in a polymer solution. The nanostructures may be dispersed or delaminated in a solvent or a solvent mixture due to the interactions of nanomaterials and the solvent molecules. However, direct mixing usually does not give homogeneous solutions because of the agglomeration of nanoparticles. Ultrasonication method can be used to enhance dispersion. Polymer chains are then adsorbed onto the delaminated or dispersed individual layers or particles. However, upon removal of the solvent, the layers or particles can reassemble to reform the stack, sandwiched in between polymer chains and forming a well-ordered intercalated nanocomposite. The preparation of exfoliated nanocomposite is very difficult by this method. This method is often used in the laboratory for the preparation of nanocomposites but industry used only when water is used as the solvent. However, this method is not well accepted for the commercial production of nanocomposites because of the requirement of large amount of solvent and phase separation of the prepared product. Further, this method is environmentally unfriendly because of the usage of large amounts of solvents.

1.7.1.2 Melt mixing

This technique employs the mechanical mixing of a polymer with nanomaterials using extrusion and injection molding above the softening point or at processing temperature of the polymer. The intercalation or exfoliation of nanomaterials takes place during melting of polymer under shear conditions. The processing conditions play significant role on the structure evaluation of such polymer nanocomposites. Generally, temperature is kept well below the decomposition temperature of all the constituents of the nanocomposites. The absence of a solvent makes direct melt intercalation an environmentally sound and an economically favorable method for industries. Thus this technique has wide application for commercial production of polymer nanocomposites.

1.7.1.3 In situ polymerization

This method is important for the preparation of polymer nanocomposites, which cannot be processed by solution or melt mixing such as insoluble and thermally unstable polymers. In this method, the nanomaterials are first dispersed in the liquid monomer or prepolymer solution before polymerization. Polymerization can be initiated either by heat or radiation or by the diffusion of a suitable initiator or catalyst inside the nanomaterials before the swelling step by the monomer. Polymerization generates long-chain polymers within the nanomaterials. The homogeneous distribution of nanomaterials can be achieved by using a high shear device because of the gradual increment of viscosity of the reaction mixture with progress of polymerization. This technique is more efficient to prepare well-exfoliated nanocomposites and has been applied to a wide range of polymer systems. This method is mostly useful for the thermosetting polymers.

1.7.2 Different SMP nanocomposites

1.7.2.1 Nanoclay-based SMP nanocomposites

Montmorillonite, attapulgite, saponite, talc and synthetic mica are usually used clay materials for the preparation of clay nanocomposites. Nanoclay improves the modulus, toughness, thermostability, barrier properties and SME of SMP [2, 59, 95–97]. The clay particles are layer structures, where each layer is composed of octahedral and tetrahedral types of structural sheets (Figure 1.6). The tetrahedral sheet is composed of silicon-oxygen tetrahedra linked to neighboring tetrahedra

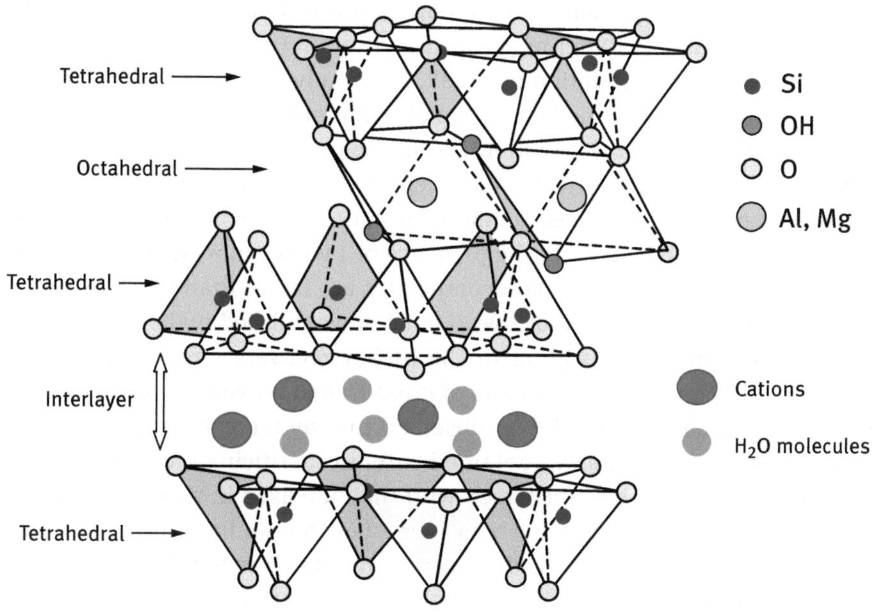

Figure 1.6: Structure of clay particle.

by sharing three corners, producing a hexagonal network. The remaining fourth corner of each tetrahedron forms a part to adjacent octahedral sheet. The octahedral sheet is usually formed from aluminum or magnesium in sixfold coordination with oxygen from the tetrahedral sheet and with hydroxyl. The two sheets together form a layer, and a number of layers may be joined in a clay crystallite structure through interlayer cations, electrostatic force, van der Waals force or by hydrogen bonding. However, most of the clay structures tend to have a negative charge, the inorganic ions such as Na and K between galleries and hold negatively charged layers together. The substitution of the inorganic ions in the galleries of the clay by alkylammonium salts or quaternary amines and/or phosphonium salts with long alkyl chains substituent possesses a better compatibility between the clay materials and hydrophobic polymer matrix. Based on morphological structures, clay-polymer nanocomposites can be classified into two categories, namely intercalated and exfoliated nanocomposites. In the intercalated structure, the polymer chains are being inserted into the silicate layered structures in a crystallographically regular manner, whereas in exfoliated structure the individual silicate layers are segregated in the polymer matrix.

The effect of nanoclay on mechanical and the shape memory properties of nanoclay/polyurethane nanocomposites has been reported by Cao and Jana [98]. The tensile strength and elongation at break found to increase with the increase in the content of nanoclay. The enhancement of tensile stress can be attributed to the increase in interfacial bonding between reactive clay and polymer chains, resulting in retardation of the molecular chains mobility. Furthermore, the quaternary ammonium ions present in clay particles effectively acted as a plasticizer, which contributes to dangling chain formation in the matrix and as a consequence increased the elongation at break. The study revealed that the shape fixity found to decrease with the increase in clay content. It was reported that the decrease in crystallinity of the polyurethane in the presence of nanoclay was the consequence of decrease in fixity. It was observed that shape recovery stress was found to increase with the increase of nanoclay content. The magnitude of recovery stress was increased by 20% in the presence of 1 wt% nanoclay. However, the recovery stresses of 3 and 5 wt% nanocomposites are lesser than those of 1 wt% case. Nanocomposites composed of 1 wt% nanoclay exhibited higher exfoliation and lower relaxation rate, resulting in the higher recovery stress. Mechanical properties and shape memory behaviors of cross-linked polyethylene/clay nanocomposites have been reported. It was observed that by increasing clay content, an increase in elastic modulus was obtained without sacrificing the elongation at break. The recovery stress was increased with the increased in nanoclay content as a result of increase in modulus. It was suggested that an increase in recovery stress in shape memory cross-linked polyethylene/clay nanocomposite renders their application as high-performance light-weight actuator with reasonable recovery strains.

Shape memory and mechanical properties of attapulgite clay/polyurethane nanocomposites have been investigated in wet and dry condition of the clay [99]. Attapulgite (another name palygorskite) is a kind of nanosized fibrous crystalline hydrated magnesium-aluminum silicate mineral [$(Mg, Al)_2Si_4O_{10}(OH)_4(H_2O)$], which can expand dramatically upon wetting. The untreated commercial attapulgite clay-based nanocomposites exhibited a significant decrease in T_g and hardness owing to the presence of moisture as well as the clay's amorphous structure and surface hydroxyl groups, resulting in plasticizing effect. The nanocomposites based on heat-treated clay exhibited significant enhancement in hardness with the increase in clay content, which is due to the homogeneous dispersion of the nanoclay in the polymer matrix and efficient interactions between nanoclay and polymer matrix. The nanocomposites showed the similar shape recovery behavior as pristine polyurethane although recovery speed was found to be slightly slower.

Shape memory and physical properties of poly(ethyl methacrylate)/sodium montmorillonite (Na-MMT) nanocomposites have been studied [100]. A macroazoinitiator (MAI) having a PEG segment was intercalated in the gallery of Na-MMT. Na-MMT intercalated with PEG exhibited high stability, as intercalated PEG cannot be replaced by organic compounds, which has high affinity toward Na-MMT. This intercalated MAI was used in the preparation of poly(ethyl methacrylate)/Na-MMT nanocomposites by means of in situ radical polymerization of ethyl methacrylate. The nanocomposites showed higher modulus and T_g than the pure polymer, as the dispersed fillers decrease the chain mobility of matrix polymer and strengthen the matrix. The nanocomposites exhibited higher rubbery plateau than the pure polymer even at low content of Na-MMT. This is because Na-MMT intercalated with a PEG segment acts as filler or physical cross-linker which hinders the slippage or the flow of polymer segments at the rubbery state. In the shape memory study, a minor increase in residual strain was observed when the repeated test cycle was increased as compared to pristine polymer. These results showed that Na-MMT intercalated with a PEG segment was well dispersed in the polymer matrix and acted as a fixed phase. Thus, nanomaterials resist the slipping and disentangling of the polymer chains which inhibits the permanent deformation with the increase of test cycle. It was observed that the residual strain found to increase in excess amount of nanomaterials content. This is because of the probability of deformation of fixed phase, Na-MMT intercalated with a PEG segment, together with a reversible phase by external force which is increased at high content of the fixed phase.

Shape memory behaviors and mechanical properties of starch/clay bionanocomposites have been investigated [101]. The nanocomposites showed enhanced elastic modulus and tensile stress above T_g up to 5 wt% of nanoclay. However, no significant enhancement of modulus and tensile stress was observed above 5 wt% of nanoclay. The nanocomposites exhibited higher recovery stress as compared to the pristine starch. It was suggested that the bionanocomposites could be promising perspectives for the design of sensors and actuators.

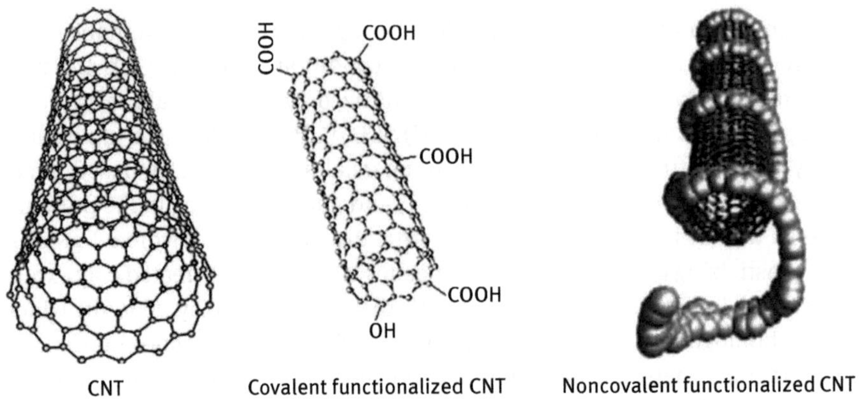

Figure 1.7: Structure of CNT.

1.7.2.2 CNT-based SMP nanocomposites

The excellent mechanical and electrical properties of CNT have taken immense attention in the fabrication of advanced SMP nanocomposites. CNTs are seamlessly rolled sheets of hexagonal array of carbon atoms with diameter ranging from a few Angstroms to several tens of nanometers (Figure 1.7). The tensile modulus and strength of CNT lie in the order of 270 GPa to 1 TPa and 11–60 GPa, respectively [102–104]. CNTs dramatically improve the storage modulus, toughness, fatigue strength and shape memory functions of SMP nanocomposites [59, 105–107]. CNT-based SMP nanocomposites show enhanced recovery stress and recovery speed as compared to the pure SMP. CNTs have an anisotropic nature enabling a percolative network at low volume fractions in the SMP matrix. The conducting SMP nanocomposites by the incorporation of CNT could be actuated via electric field. Moreover, the properties of nanocomposites depend on dispersion, orientation and distribution of CNT in the polymer matrix. However, the preparation of CNT nanocomposites with well distribution of CNT in the polymer matrix is very challenging, because CNT is susceptible to form agglomerates that are thermodynamically stabilized by van der Waals forces and numerous π–π interactions between the nanotubes. Therefore, modification of CNT is needed for the homogenous distribution of CNT in the polymer matrix to ensure strong CNT–polymer interfacial interaction by means of good load transfer ability. Various methods such as noncovalent and covalent modification have been employed to modify CNT to augment the dispersion as well as load transfer ability. Noncovalent modification is primarily based on the adsorption of functional molecules on the surface of the nanotubes through different interactions such as hydrophobic interaction, van der Waals forces, electrostatic forces and π–π interactions. In the noncovalent modification, the intrinsic property of the CNT is preserved. In the covalent modification, the functional groups covalently attach onto the side wall or ends of the nanotubes. The end caps of the CNT have the highest reactivity owing to its higher pyrimidization angle, while the walls of the CNT have lower reactivity because of its lower pyrimidization angles. However, in covalent modification significant changes in physical and mechanical properties of CNT take place, because the breaking of carbon–carbon bonds induce defects and introduce an insulating layer on the surface of CNT during covalent functionalization. However, most of the covalent modifications are performed by using concentrated HNO_3 and H_2SO_4 or mixture of concentrated HNO_3/H_2SO_4, which are harmful to the human body.

Epoxy resin/CNT-based SMP nanocomposites have been fabricated [108]. The recovery stress of the nanocomposites was investigated in deformation at different temperatures such as 25, 100 and 150 °C. The nanocomposites exhibited higher recovery stress than the pure SMP due to the strain storage of CNT in the programming process. Moreover, the nanocomposites deformed at 100 °C showed higher recovery stress than the other deformation temperature. It was suggested that the stable recovery stress depends on the shape fixity ratio and viscoelastic properties during the programming process, which is dominated by the deformation temperature. The extent of CNT deformation of nanocomposites at 25 °C is small, resulting in the low CNT reinforcement efficiency for the stable recovery stress. Further at high deformation temperature such as 150 °C, the specimen appeared in rubbery state and the extent of CNT deformation was limited by the low stiffness of the polymer chains. The nanocomposites deformed at 100 °C, possess good combination of suitable stiffness, and conformational motion ability of the polymer chains resulted in a relatively larger extent of CNT deformation, which leads to higher CNT reinforcement efficiency for the stable recovery stress. Nanocomposites deformed at 100 °C showed higher recovery speed as compared to the deformation temperature at 25 and 150 °C. This can be attributed to the extent of strain energy stored in CNT during deformation at 100 °C was more than those at 25 and 150 °C. The nanocomposites deformed at 25 °C, the interface failed, which prevented the recovery of the original shape, and as a consequence lower the recovery speed.

Mechanical and shape memory properties of pristine CNT and oxidized CNT-incorporated low-temperature actuated polyurethane nanocomposites have been investigated [109]. Both the nanocomposites exhibited higher tensile strength and elongation at break than the neat polyurethanes. Moreover, oxidized CNT-based nanocomposites showed improved tensile strength than the pristine CNT-based one because of the enhanced interfacial bonding between the oxidized CNT and the polymer matrix. The nanocomposites exhibited higher storage modulus than the pristine polyurethane in glass state, which is attributed to the stiffness effect of the CNT. However, the nanocomposites showed lower storage modulus than the neat polyurethane above T_g, indicating that rubbery modulus of the nanocomposites is lower than the pure polyurethane. However, oxidized CNT-based nanocomposites possess lesser rubbery modulus than that of pristine CNT one. The decrease in rubbery modulus of the nanocomposites may be due to the disturbance of diisocyanate segments caused by incorporation of the CNT and the disturbance was more effective in case of oxidized CNT due to the well dispersion of oxidized CNT. From the stress–strain curves, it was demonstrated that the oxidized CNT-based nanocomposites have higher recovery force than the pristine CNT-based one. As the oxidized CNT-based nanocomposites exhibited higher integral area under the curve as compared to pristine CNT-based one. The area under the curve denotes the strain energy stored during stretching and it drives strain recovery upon release of stress in its rubbery state. Neat polyurethane and its nanocomposites exhibited excellent shape fixity above 98% for both first and second tensile cycles. Pure polyurethane and its nanocomposites showed shape recovery ratio above 85% for the first tensile cycles while above 98% for the second tensile cycle. The lower recovery in the first cycle is attributed to the segment-chain orientation and relaxation effects. It was suggested that these kinds of nanocomposites have potential applications for controlling tags or proof marks in the area of frozen food.

Kalita and Karak [110] studied the mechanical properties, thermal stability and shape memory behaviors of biobased polyurethane/multiwalled carbon nanotube (MWCNT) nanocomposites. Pristine MWCNTs were modified through noncovalent approach using polyoxyethylene octyl phenyl ether (Triton X-100) for better interfacial interactions. The nanocomposites exhibited enhanced tensile strength, impact resistance and scratch resistance than the pure polyurethane. Furthermore, the nanocomposites showed higher thermal stability than that of pure polyurethane. This can be attributed to the homogeneous distribution of MWCNT and increased interfacial interaction like physical cross-linking between the MWCNT and polymer matrix. Therefore, macromolecular chains are immobilized on the surface of MWCNT that hinders the thermal motion of the macromolecular chains in the nanocomposites. In addition, the MWCNT served as physical barrier, so the volatile components formed during the thermal decomposition process have to overcome longer zigzag path to escape from the matrix. Furthermore, MWCNTs are the most thermostable and thermal conducting materials, as a consequence, enhanced thermal conductivity of the nanocomposites can assist well dissipation of the thermal energy and thus enhanced the thermal stability. The pristine polyurethane and its nanocomposites showed excellent shape fixity (100%) and can be stored for a long period at room temperature, which implied that micro-Brownian motions of the molecular chains in amorphous network were frozen during vitrification. The nanocomposites showed faster shape recovery than that of pure polyurethane. This can be attributed to the enhanced elastic strain energy in the nanocomposites due to the homogeneous distribution of the MWCNT and the increased physical cross-linking between the MWCNT and the polymer matrix. Moreover, the high rate of heat transfer in the nanocomposites is due to the fact that the presence of conducting MWCNT may also be responsible for the faster shape recovery. Kalita and Karak [111] also studied the mechanical and shape memory behaviors of Fe_3O_4 nanoparticle decorated with MWCNT (Fe_3O_4-MWCNT)/hyperbranched thermosetting polyurethanes. Similarly the nanocomposites showed enhanced tensile strength, impact strength and thermal stability as compared to pristine polyurethane. The enhanced tensile strength can be attributed to the well dispersion of Fe_3O_4-MWCNT and enhanced various interactions among the Fe_3O_4-MWCNT, epoxy and the polymer matrix. As a consequence, molecular chains are restricted on the surface of the MWCNT resulting in decrease in chains slippage during the application of loading. However, the elongation at break decreased with the increase of the content of Fe_3O_4-MWCNT in nanocomposites. This is because of the restricted movement of macromolecular chains due to the presence of various physicochemical interactions in the nanocomposites. The hyperbranched polyurethane and its nanocomposites exhibited excellent shape recovery under the microwave stimulus. No significant change of shape recovery of the nanocomposites was found over the repeated cycles (ten) of test. The shape recovery is due to the generation of induced heating in the samples. When the microwave is irradiated on the sample, the dipole moment oscillates to align with the external electric field; as a consequence, heat is generated because of the molecular friction and collisions. The shape recovery effect was observed when the induced heat is close to the T_{trans} and found to be 45 °C. The shape recovery speed was found to increase with the increase of Fe_3O_4–MWCNT, which may be due to the increased microwave absorption characteristic of the nanohybrid system as both Fe_3O_4 and MWCNT are microwave absorbing materials.

Jin Yoo et al. [112] reported the mechanical and electroactive SME of polyurethane-MWCNT nanocomposites. The tensile strength and modulus were found to enhance after incorporation of MWCNT without sacrificing the elongation at break of the material. This is because of the homogeneous dispersion of MWCNT in the matrix as well as efficient interfacial interactions between MWCNT and polyurethane matrix. Thus the interface acts as a vital role in reinforcing the nanocomposites by effectively transferring the stress between MWCNT and polyurethane matrix. It was pointed out that mechanical properties also strongly depend on microstructure and crystallinity of the materials. The nanocomposites that possess conductivity 0.28 S/cm particularly exhibited electroactive SME. This is because the conductivity was sufficient to make the sample heated above T_{trans} of 36–42 °C due to melting of PCL soft segment domain. The samples recovered their original shape in 40 s with bending mode when an electric field of 50 V was applied.

1.7.2.3 CNF-based SMP nanocomposites

CNFs significantly improve the mechanical and shape memory properties of SMP because of its high elastic modulus, high aspect ratio and good electrical and thermal conductivity [113–115]. CNFs have strong influence on crystallinity, cross-linking density and conductivity of SMP that lead to effect on SME of SMP. By incorporation of CNF, SMP nanocomposites can be actuated by multiple stimuli. Furthermore, CNF have been applied as promising materials in fields of energy conversion and storage, self-sensing devices and smart materials. CNFs have diameters of 10–100 nm and length of 10–100 μm. CNFs are mostly synthesized by two methods, namely catalytically vapor deposition growth and electrospinning. In the catalytically vapor deposition growth method, metals or alloys (Fe, Cr, Co and Ni) are used as catalyst, and carbon monoxide, ethene, ethyne and synthesis gas (H_2/CO) are used as a carbon source at temperature range of 700–1,200 K [116, 117]. In the electrospinning method, polymer nanofibers are prepared as precursor of CNF, where poly(acrylonitrile), poly(vinyl alcohol) and phenolic resin are used as carbon source for the preparation of polymer solution to spinning. Once the polymer nanofibers have been successfully fabricated, a heat treatment of the polymer nanofibers will be applied to carbonize the polymer nanofibers to form CNF. The properties such as morphology, crystallinity and diameters of CNF depend on the parameters of heat treatment process. The performances of the CNF-based nanocomposites mainly depend on the dispersion of the CNF in the polymer matrix. The chemical surface treatment of CNF is an efficient technique to enhance their dispersion in the polymer matrix. Usually the surface treatment is done by oxidizing the CNF surface reacting in a mixture of sulfuric and nitric acid at various temperatures.

The mechanical and electrical actuation shape memory behaviors of epoxy/CNF nanocomposites have been investigated by Luo and Mather [118]. The nanocomposites exhibited higher tensile strength and rubbery modulus than the pristine epoxy. This renders the nanocomposites suitable for applications where large recovery stresses are desired. The nanocomposites showed excellent shape recovery under the application of constant DC voltage. This can be attributed to the well dispersion of CNF and the generation of efficient percolating conductive network. They studied the shape recovery under different DC voltages at 10, 15 and 20 V. The recovery speed increased with the increase in applied voltage. Both the induction times and recovery times found to decay exponentially with applied voltage. Furthermore, the induction times can be used to estimate the initial heating rate. The initial heating rate ($\mathrm{d}T/\mathrm{d}t$) can be calculated with the equation: $\mathrm{d}T/\mathrm{d}t = T_g - T_e/T_i$, where T_e and T_i are the environmental temperature (here room temperature) and induction time, respectively. The calculated initial heating rates were found to be 25.7, 11.09 and 4.73 °C/s for 20, 15 and 10 V, respectively. The nanocomposites exhibited quite high heating rates and cannot be easily achieved by conventional external heating, even for SMP with enhanced thermal conductivity. Thus the nanocomposites not only provided high electrical conductivity but also significantly improved heat transfer and together resulting in high actuation speed. These materials have great potential in applications encompassing actuators, sensors and deployable devices.

Electroactive SME of vapor-grown carbon nanofiber (VGCF) reinforced biobased polyester has been studied [119]. VGCF significantly improved the mechanical properties of the nanocomposites. The neat biobased polyester showed tensile strength of 0.13 MPa, while nanocomposites with 5.0 vol% VGCF showed 1.52 MPa at temperature (T_m + 30 °C). The nanocomposites exhibited increase in elongation at break with initial incorporation of VGCF (up to 5.0 vol%) and then reduced at higher loadings. This is due to the fact that debonding and mobility of VGCF in the polymer matrix facilitates the energy dissipation, resulting in the toughening of the composites at low content of nanofiber. At higher content of VGCF facilitates the formation of a VGCF network that restricts the mobility of polymer chains, leading to the brittle response. The nanocomposites exhibited higher storage modulus as compared to the pristine polyester. This is due to the homogenous dispersion of VGCF and the fairly strong interfacial interaction between VGCF and polyester which afford efficient load transferring from polyester matrix to VGCF. The pristine biobased polyester and its nanocomposites possess high E' ratio below T_c and above T_m (>100), signifying that the synthesized biobased polyester and its nanocomposites favor the deformation above T_m and fixing below T_c. This implies the nanocomposites as new SMP with excellent SME. The shape recovery of the materials was studied in different voltages such as 10, 20, 30 and 40 V, and the shape recovery speed was found to increase with the increase in the voltage. It can be seen that the nanocomposites with higher content of VGCF (11.6 vol%) showed faster recovery within 20 s with an applied voltage of 10 V. This can be attributed to the efficient inductive joule heating effect due to the sufficient conductivity (14.4 S/m) of the materials. However, nanocomposites with 7.3 vol% of VGCF have not shown electroactive actuation below 50 V due to the less conductivity (1.7 S/m) of the materials. The amount of the generated Joule heat (Q) depends on the applied voltage (V), resistance (R) of the sample and time (t) with the equation of $Q = (V^2t)/R$. According to the equation, a higher applied voltage and lower resistance, i.e., higher conductivity of the material, will generate more heat in unit time. Several fascinating attributes, including biobased nature, good mechanical properties and excellent electroactive SME, render these nanocomposites as good candidates for electrically actuated sensors and actuators.

Jimenez and Jana [89] demonstrated the mechanical properties and SME of CNF-based TPU nanocomposites. It was found that neat polyurethane could be stretched only up to 76% at 60 °C, whereas CNF composites could be easily stretched to 100% of their length. The nanocomposites exhibited slightly lower storage modulus than the neat polyurethane at their glassy state due to the reduction of PCL crystallinity. However, nanocomposites exhibited higher storage modulus at their rubbery state due to the prominent reinforcing effect of the CNF. The glass transition peak for pure polyurethane was broad and showed low tan δ values, indicating that less amorphous polymer chains participated in glass transition and that PCL-diol crystallinity was higher. Nanocomposites showed higher tan δ values, indicating the increase of the amorphous polymer chains in the composites. It was observed that shape fixity of the nanocomposites decreased with the increase of CNF content. This can be attributed to the decrease in crystallinity of the PCL in nanocomposites. Nanocomposites exhibited higher recovery stress as compared to the pristine polyurethane. The surface temperature of the samples with 5 wt% of CNF reached a value more than 50 °C within 45 s with the application of 300 V. It was observed that the nanocomposites with 5 wt% of CNF exhibited shape recovery within 60 s under the application of 300 V. The study revealed that both PCL-diol crystallinity and the presence of CNF possess significant role to the tune of shape fixity and recovery stress in electrically actuated shape memory behavior.

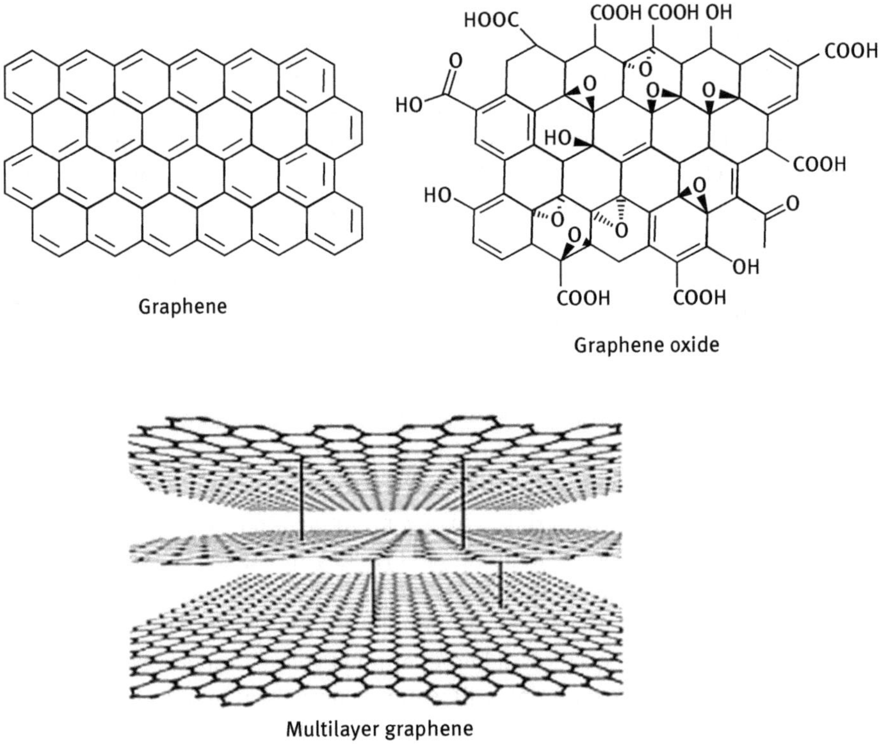

Figure 1.8: Structure of graphene.

1.7.2.4 Graphene-based SMP nanocomposites

Graphene has attracted immense interest because it dramatically improves the physical properties and SME of SMP. Graphene is a two-dimensional atomic layer of sp^2 hybridized carbon atoms arranged in a honeycomb structure, which exhibits extraordinary mechanical strength (Young's modulus ≈ 1,086 GPa), thermal conductivity (5,300 W/mK), electrical conductivity (6,000 S/cm) and high specific surface area (2,600 m^2/g) (Figure 1.8) [120–123]. Graphene has promising applications in supercapacitors, batteries and sensors. Graphene can be synthesized by bottom-up and top-down approaches [124, 125]. In the bottom-up approaches, graphene is synthesized in different methods such as chemical vapor deposition, arc discharge, reduction of CO and epitaxial growth on SiC. In the top-down methods, graphene is synthesized by separation or exfoliation of graphite and graphite derivatives such as graphite oxide and graphite fluoride. For large-scale synthesis, first graphite oxide is prepared from natural graphite powder or flakes, converted to graphene oxide by ultrasonication followed by reduced graphene. However, large-scale production processes 99% of the material produce as multilayer graphene nanoplatelets and only about 1% as true monolayer graphene sheets. SMP nanocomposite inclusion of graphene as nanofillers substantially improve the shape memory properties such as recovery stress, response time and recovery ratio, and confers a multistimuli-responsive SMP such as infrared radiation, electric and magnetic fields [126–129]. In addition, graphene remarkably improves the modulus, tensile stress, fracture toughness, fracture energy and fatigue resistance of SMP. However, homogeneous dispersion and efficient interfacial interactions of graphene and polymer matrix are difficult because of severe aggregation due to the large surface area and strong van der Waals force among graphene sheets. The interfacial interactions between the graphene and the polymer matrix can be improved by functionalizing the graphene surfaces. Noncovalent and covalent modifications of graphene surfaces are extensively explored to improve its dispersion. The precursor of graphene, graphene oxide, has plentiful functional groups on the surface such as hydroxyl, epoxy, ketone and carboxyl, which provides the reactive site for covalent functionalization. Covalent functionalizations of graphene are shown in Figure 1.9. In the noncovalent modification, the functional molecules can be

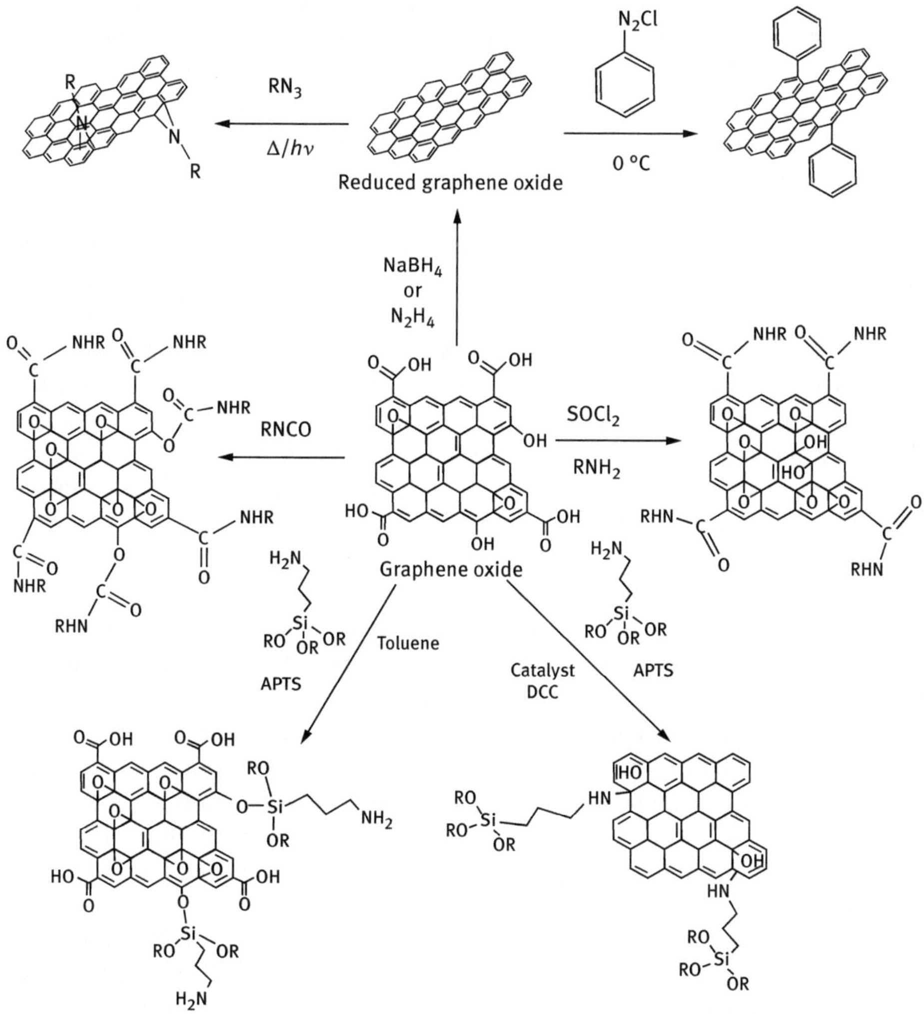

Figure 1.9: Covalent functionalization of graphene.

assembled on the surface of graphene sheet through van der Waals interactions, π–π interactions, hydrogen bonding and electrostatic interactions.

Thermal, mechanical and SME of graphene/polyimide nanocomposites have been investigated [126]. Exfoliated graphene nanosheets were synthesized by the rapid high-temperature thermal decomposition of graphene oxide. To enhance the compatibility, graphene surfaces were modified with *p*-phenylene trimethoxyaminosilane. The storage modulus of the nanocomposites increased with the increase in graphene content (0.1–4 wt%) below T_g. This can be attributed to the reinforcement effects of graphene in the polyimide resin. The increase in the rubbery state modulus was not significant at lower content and only observed at higher content (2 and 4 wt%) of graphene in nanocomposites. The nanocomposites exhibited tan δ peak at 212 °C and have no significant effect of increase in graphene content. However, the area under damping peak (tan δ) was found to decrease with increase in graphene loading. This can be due to the weak attractive forces between the graphene and the polyimide resin matrix, which are insufficient to hinder segmental motion of the polymer chains. The thermal stability was increased with the increase of graphene content. Both pristine polyimide and polyimide nanocomposites performed shape memory behavior with a switching temperature of 230 °C. The recovery rate of the polyimide enhanced with the addition of graphene.

Jung et al. [130] reported the fabrication of transparent, tough and conductive shape memory polyurethane films by incorporating graphene. Graphene was synthesized by solution phase exfoliation of expandable graphite. A small amount of graphene (0.1 wt%) enabled the modulus and tensile strength of the nanocomposite film to be enhanced from 5 to 62.9 MPa and from 15.6 to 37.4 MPa, respectively. This can be attributed to the reinforcing effect of the graphene via various interactions such as hydrogen bonding, dipole–dipole and induced dipole–dipole interactions in the nanocomposites. A large enhancement of thermal conductivity of the nanocomposite film was observed, which could be attributed to the homogeneous dispersion of the graphene sheets within the polymer matrix and strong interfacial interactions between graphene sheets and polymer backbone. The nanocomposites exhibited higher recovery force and shape recovery ratio than the pristine polyurethane. This is due to the homogeneous dispersion of graphene in the polymer matrix and efficient interaction between graphene and polyurethane matrix.

Infrared light-actuated poly(L-lactic acid) (PLLA)/graphene SMP nanocomposites have been reported [131]. The nanocomposites exhibited higher storage modulus than the pristine PLLA due to the reinforcing effect of the graphene. However, the storage modulus decreased above 4 wt% of graphene content, which can be due to the aggregation of nanofillers and lowering the stiffening effect of graphene. The tan δ value found to decrease than the pristine PLLA due to decrease in the energy dissipation as a result of the presence of elastic graphene nanomaterial. The pristine PLLA exhibited SME due to crystalline and amorphous phases of PLLA acting as fixed and reversible phases, respectively. The nanocomposites exhibited higher shape fixity and shape recovery than that of neat PLLA. This may be attributed to the enrichment of crystallinity and thermomechanical properties of nanocomposites enabling more stable fixing phase leading to better shape fixation and recovery. Furthermore, graphene enhances heat conductivity as a result of more efficient heat transfer during cooling and heating processes leading to faster response with less dissipation of energy and provide higher shape fixity and recovery in the nanocomposites. However, the shape memory parameters achieved decreasing trend beyond 1 wt% of graphene content, which can be attributed to the agglomeration and crystallinity reduction leading to weakening of fixed phase and lowering SME. The shape recovery showed decreasing trend with increasing cycle number in the shape memory test, due to irreversible processes such as slippage of crystalline domains, plastic deformation of PLLA chains and cleavage of some physical linkage. The nanocomposites also exhibited infrared light-activated SME. Graphene enhanced infrared light absorption of PLLA remarkably and transforming infrared light into heat corresponds to switching temperature, which leads to faster and higher recovery of original shape.

SME of polyester/CNF–graphene hybrid nanocomposites has been investigated [132]. CNF–graphene hybrid was prepared by direct reduction of graphene oxide in NMP with the assisted dispersion of CNF. CNF acted as nanoparticles like stabilizer to avoid the restacking of graphene by adsorbing CNF onto graphene sheets. This type of nanoparticle-based modification/stabilization of graphene has several advantages over covalent and noncovalent modifications. In the covalent modification intrinsic properties of the graphene is damaged, whereas in the noncovalent modification though the intrinsic properties are preserved, the residual stabilizer may weaken the mechanical properties of the composites. The CNF–graphene hybrid-based nanocomposites exhibited lower electrical percolation threshold and higher electrical conductivity than the CNF-based nanocomposites. This can be attributed based on the three factors: (1) the addition of graphene for the dispersion of CNF is significantly improved; (2) the large specific surface area and homogeneous dispersion of graphene encourage the formation of conductive pathway; (3) due to the strong interactions, one-dimensional CNF and two-dimensional graphene cooperate, bridging adjacent graphene through a "conductive bridge" of CNF and forming multiple pathways in the composites. The CNF–graphene hybrid-based nanocomposites exhibited higher modulus and ultimate stress than the CNF-based nanocomposites. This can be attributed to the improved dispersion and interfacial interactions of CNF–graphene hybrid and polymer matrix with the addition of graphene. Moreover, the elongation at the break of nanocomposites was enhanced with initial incorporation of fillers and then decreased at higher loadings. Due to the incorporation of fillers, the load can be transferred onto fillers to dissipate the load, and the mobility of the fillers in the polymer matrix can also facilitate the energy dissipation, which leads to the toughening of the nanocomposites and increasing the elongation at break. The nanocomposites composed of 10.3 vol% of CNF–graphene hybrid performed shape recovery within 60 s under the application of 30 V. However, nanocomposites composed of 10.3 vol% of CNF cannot be effectively actuated under the

30 V. This is because that the electrical conductivity of CNF nanocomposites (4.0 S/m) is much lower than that of CNF–graphene hybrid nanocomposites (11.5 S/m), as a result, less Joule heat is generated and cannot efficiently induce the shape recovery.

Han and Chun [133] studied the SME of polyurethane nanocomposites incorporation of functionalized graphene. Graphene oxide nanosheets were prepared by the oxidation of graphite powder using a modified Hummers method. Then partial reduction of graphene oxide with sodium borohydride in surfactants sodium dodecylbenzenesulfonate followed by functionalization with 2-(4-aminophenyl)ethanol through a diazonium reaction, as a consequence specific functional groups were successfully introduced onto the surface of nanolayered graphene. The tensile strength of nanocomposites increased to 136% at 0.5% loading with a strain decreased to 93%. This result implied that the functionalized graphenes were well dispersed in the polyurethane matrix through various interactions between the oxygen containing groups and the urethane linkages. Furthermore, stronger reinforcing effect of nanocomposites is due to the covalent cross-links between the terminal hydroxyl groups of functionalized graphene and the NCO end groups of the polyurethane. The nanocomposites showed higher storage and rubbery modulus than the pristine polyurethane due to the reinforcing and cross-linking effects of functionalized graphene. Pure polyurethane exhibited poor shape fixity and shape recovery, which is caused by the low rubbery modulus and rapid chain relaxation. With the loading of functionalized graphene, the shape fixity and the shape recovery increased, due to the larger rubbery modulus and the longer relaxation time. A much better SME was the result of the introduction of a network structure built with the covalent cross-linking of polyurethane and functionalized graphene. The cyclic hysteresis loss resulting from chain slippages and breakages during the repeated loading and unloading was reduced by the incorporation of functionalized graphene, and further reduced by the cross-linked structure. The hysteresis loss was found to be 2% after four cycles, which sufficiently satisfied the essential conditions for various shape memory applications.

1.7.2.5 Other nanoparticles (SiC, POSS)-based SMP nanocomposites

Silicon carbide (SiC) nanofillers significantly improve the elastic modulus and shape recovery stress of SMP [134–136]. SiC nanoparticles possess high thermal conductivity, high thermal stability, good wear resistance, a small thermal expansion coefficient and strong microwave absorbing material [94, 137]. It has beneficial impact on the micro-hardness, strain recovery and recovery force of SMP. SiC is a binary compound consisting of Si and C, where each Si atom is bonded to four C atoms and each C atom is bonded to four Si atoms. Various techniques such as sol–gel process, chemical vapor deposition, laser gas-phase pyrolysis and carbothermal reduction have been used to synthesize nano-SiC [138]. Carbothermal reduction method has been considered as the most effective and economical to synthesize nano-SiC using various carbon and silicon sources.

The storage and release of internal stress in nanofiller SiC-reinforced epoxy SMP matrix have been studied by Gall et al. [139]. The compressive stress–strain response of the nanocomposite was studied at a deformation temperature of 25 °C. The material exhibited an elastic response up to approximately 50 MPa followed by yield and inelastic flow up to 50% strain. The residual inelastic strain after complete unloading is found to be approximately 35%. The materials exhibited complete shape recovery at a temperature of 120 °C. They elucidated the cause of shape recovery of deformed fixed samples under compression loading. After compressing, the sample achieves a critical stress leading to a relative chain slippage and generates a metastable structure. The metastable structure in the amorphous polymer has both stored strain energy and enhanced chain organization (lower entropy). Upon unloading, the physical entanglements (temporary bonds) give a storage mechanism for macroscopic stresses in the form of local internal stresses resulting in a changed state of entropy. After subsequent heating, the bonds are weakened and the local internal stress field and low entropy state forced individual chains to return to their original positions, driving the shape recovery. Gall et al. [140] also reported the microhardness, elastic modulus and shape memory behaviors of nano-SiC-based epoxy SMP nanocomposites. They prepared the various weight fractions of 10%, 20%, 30% and 40% of SiC-based SMP nanocomposites by microcasting. The microhardness and elastic modulus of the nanocomposites were found to increase by approximately a factor of 3 with the incorporation of 40 wt% SiC into the epoxy resin. They reported that the unconstrained strain recoverability of the nanocomposites depended on the weight fraction of SiC. The recoverability of the nanocomposites was found to be perfect for SiC weight fractions below 40% for 180° bend test. Constrained bending recovery force in the nanocomposites was increased by 50% with the incorporation of 20 wt% SiC. The nanoparticles imparted to store internal elastic strain energy during loading and freezing. The same mechanism that limits the generation of large recoverable strain provides higher recoverable force levels.

Microwave-induced SME of poly(vinyl alcohol)/SiC nanocomposites have been investigated by Du et al. [137]. They modified SiC by silane coupling agent 3-triethoxysilylpropylamine to improve the dispersion. SiC is a high dielectric loss material; it can strongly absorb microwave so as to be heated to high temperature within short time. The nanocomposites exhibited good shape recovery behavior for more than 90%. The recovery time was found to decrease with the increase of SiC content. The nanocomposite with 1.5% SiC recovered its original shape within 1 min under 300 W microwave irradiation, whereas 0.03% SiC once recovered at about 2 min. This is due to the more heat generation capacity because of the presence of more amount of SiC. However, the materials did not show SME, when the microwave output power is less than 200 W. Likitaporn et al. [136] also observed the microwave actuating ability of SiC whisker nanoparticle-filled benzoxazine epoxy polymer alloy. Shape recovery time under microwave actuating ability was substantially shortened to be in a range of 3–5 min. The storage modulus (5.1–8.8 GPa) and recovery force (3.4–11.2 MPa) were found to enhance in the nanocomposites with 20 wt% of SiC. This is due to the intrinsically

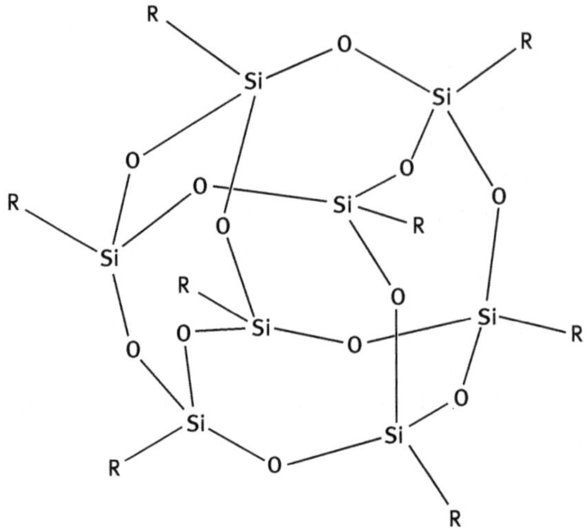

Figure 1.10: Molecular structure of POSS.

high elastic modulus of the nanofiller, which could enhance elastic modulus and stiffen the stable network of polymer nanocomposites. It was suggested that SiC whisker nanoparticle-filled SMP nanocomposites would be highly attractive to be used in multiple applications, that is, either thermal actuator or microwave actuator.

Polyhedral oligomeric silsesquioxanes (POSSs) have received extensive attention because they remarkably enhanced the stiffness, hardness, oxidation resistance, flame resistance and SME of SMP [141–145]. POSSs are three-dimensional inorganic–organic hybrids with the generic formula $(RSiO_{1.5})_n$, where R may be a hydrogen atom or an organic functional group such as alkyl, alkylene, acrylate, hydroxyl or epoxide unit (Figure 1.10) [146, 147]. It contains the rigid inorganic core (diameter: ≈0.5 nm) of silicon and oxygen with eight organic functional groups, whose diameter is in the range of 1–3 nm. POSS can be synthesized by hydrolytic condensation of trifunctional monomers $RSiX_3$, where R is a chemically stable organic substituent (methyl, vinyl, aryl or H) and X is a highly reactive substituent such as Cl or alkoxy. Inclusion of POSS macromers within polymeric architectures allows both the inorganic and organic phases to interact at the molecular level and offer the high-performance material. POSS macromers can be incorporated into polymer matrix by blending, copolymerization or grafting to prepare a variety of nanocomposites. Typically seven corners of POSS are occupied by identical R groups, which enhance the degree of compatibility of the POSS with the polymer matrix, while the remaining position is occupied by an organic reactive group, which gives the site for incorporation into a polymer backbone. POSSs have high propensity to self-assemble forming nanocrystals (dimension: ≈ 10–100 nm) and this inherent property leads to improvement of physical properties of POSS nanocomposites. The crystallization behavior and the ability to form chemical and physical crosslinks of POSS play a prominent role on the SME of POSS nanocomposites. A synthetic approach of POSS-PCL/polyurethane SMP nanocomposite is shown in Figure 1.11.

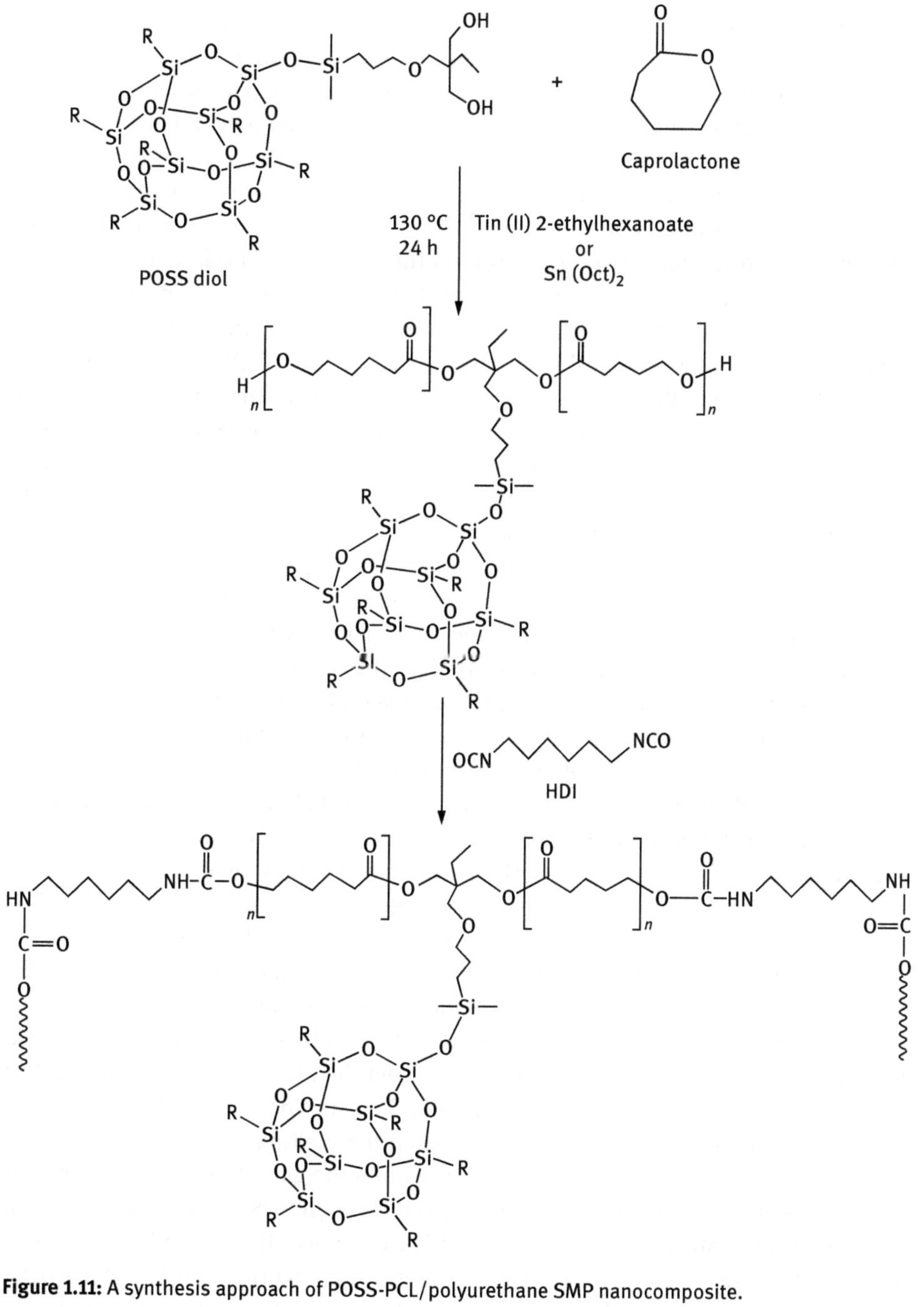

Figure 1.11: A synthesis approach of POSS-PCL/polyurethane SMP nanocomposite.

SME and mechanical properties of polyurethane/POSS nanocomposites have been investigated [145]. The nanocomposites were prepared using synthesized octa(3-hydroxypropyl) POSS (POSS-$(OH)_8$) by solution blending technique. The nanocomposites exhibited higher loss modulus than the pristine polyurethane, which is due to the constraint of the molecular chains on the POSS. The position of tan δ peak of nanocomposites shifted toward a higher value as compared to pristine polyurethane. The shift signified that the molecular mobility of amorphous polymer chains was restricted in the nanocomposites due to the presence of POSS-$(OH)_8$ and more hydrogen bonds formed between urethane linkages of polyurethane chains and hydroxyl groups on the POSS-$(OH)_8$ surfaces. The tensile strength, elongation at break and elastic modulus were found to enhance when the POSS-$(OH)_8$ amount was less than 3 wt% and gradually decreased when the amount was higher than 5 wt%, which could be attributed to the poor dispersion and agglomeration of POSS-$(OH)_8$ particles in the polymer matrix. The increase in the tensile stress with increase in elongation provides the increase in integral area under the curve, which is essential in SMP because the area is the strain energy stored during stretching and it facilitates strain recovery upon unloading of stress in its rubbery state. The nanocomposites showed the higher shape recovery ratio than the pristine polyurethane at the actuation temperature of 0 °C. Moreover, nanocomposites comprising 3 wt% POSS-$(OH)_8$ displayed highest shape recovery and faster recovery than the other compositions, which was in good agreement with the increased integral area under the stress–strain curve. The results indicated that the nanocomposites have emerging applications for controlling tags or proof marks in the area of low-temperature storage.

Knight et al. [148] studied the shape memory behaviors of poly(lactide-co-glycolide) (PLGA)/POSS nanocomposites. PLGA oligomers were synthesized to feature a hybrid organic–inorganic moiety, POSS, incorporated at the center of each chain by using it as the ring-opening initiator and end-capping with vinyl groups followed by photocured into networks with POSS content (10–41 wt%). The crystallinity of the composites increased with the increase of POSS content. Incorporation of the POSS moiety into the backbone of PLGA renders increase in the tensile modulus both above and below the T_g. The nanocomposites exhibited two rubbery plateaus due to the two thermal transitions such as a glass transition from the PLGA covalently cross-linked network and a melting transition from the POSS crystalline regions. The best shape recovery percentages (~100%) were found when the melting transition was used as T_{trans} while the best shape fixity was observed when the glass transition was used. It was reported that strain fixing can be achieved through POSS phase crystallization or PLGA phase vitrification. Knight et al. [149] also studied the mechanical properties and SME of polyurethane/POSS nanocomposites. POSS-incorporated TPUs have been synthesized by two step techniques. The initial step, synthesis of the lactide-based soft block, allowed for manipulation of T_g via controlled PEG inclusion followed by synthesis of polyurethane reacting with lysine-derived diisocyanate and POSS diol as the hard block. The crystallinity and rigidity of the material was increased with the increase of POSS incorporation level in the polyurethane hard block. Rattinger et al. [150] also observed POSS moieties assembled into higher ordered lamellar, cylindrical or cubic nanophase crystalline structures, depending on POSS inclusion level in polyurethane matrix as detected by small-angle X-ray scattering (SAXS). Furthermore, Fu et al. [151] also observe that the POSS molecules appear to form nanoscale crystals in the hard segment domains of polyurethane. It was observed that all POSS-polyurethanes exhibited storage tensile modulus greater than 2 GPa at glassy state, whereas this modulus dramatically dropped to a value of about 10 MPa upon heating above T_g. All the POSS-polyurethane nanocomposites showed quite broad rubbery plateau of modulus while the pristine polyurethane softened continuously upon heating above T_g and displayed the absence of any rubbery plateau. It can be attributed that the POSS hard block crystallization provided the physical cross-linking required for elastic mechanical behavior. The absence of rubbery plateau in pristine polyurethane indicated the poor SME due to the lack of a recovery driving force.

Shape memory performance of star-shaped POSS-PCL polyurethanes has been investigated [152]. First, star-shaped POSS-based multiarms PCL with different arm lengths were synthesized by ring-opening polymerization of ε-caprolactone followed by reacting with hexamethylene diisocyanate. At higher POSS content, a depression in PCL crystallinity was observed, which could be attributed to the effect of the positional restriction imposed onto the PCL chains by their covalent attachment to the POSS core. At higher POSS content, the materials showed enhanced shape fixities, recoverabilities and stress storage capacities. The causes of reduction of the shape memory performance at lower POSS content (higher PCL soft segment content) were a decrease in cross-link density and enhanced chain flexibility/molecular mobility. The materials showed significant cycles' averaged ($N = 2–5$) shape fixities and strain recoverabilities of 98% and stress recoverabilities close to 100%.

References

[1] Hager MD, Bode S, Weber C, Schubert US. Shape memory polymers: past, present and future developments. Prog Polym Sci 2015;49–50:3–33.

[2] Meng H, Li G. A review of stimuli-responsive shape memory polymer composites. Polymer 2013;54(9):2199–221.

[3] Pilate F, Toncheva A, Dubois P, Raquez JM. Shape-memory polymers for multiple applications in the materials world. Eur Polym J 2016;80:268–94.

[4] Zhao Q, Qi HJ, Xie T. Recent progress in shape memory polymer: new behavior, enabling materials, and mechanistic understanding. Prog Polym Sci 2015;49–50:79–120.

[5] Chen JK, Chang CJ. Fabrications and applications of stimulus-responsive polymer films and patterns on surfaces: a review. Materials (Basel). 2014;7(2):805–75.

[6] Xie T. Recent advances in polymer shape memory. Polymer 2011;52(22):4985–5000.

[7] Ratna D, Karger-Kocsis J. Recent advances in shape memory polymers and composites: a review. J Mater Sci. 2008;43(1):254–69.

[8] Leng J, Lan X, Liu Y, Du S. Shape-memory polymers and their composites: stimulus methods and applications. Prog Mater Sci 2011;56(7):1077–135.
[9] Huang WM, Yang B, Zhao Y, Ding Z. Thermo-moisture responsive polyurethane shape-memory polymer and composites: a review. J Mater Chem 2010;20(17):3367–81.
[10] Kunzelman J, Chung T, Mather PT, Weder C. Shape memory polymers with built-in threshold temperature sensors. J Mater Chem 2008;18(10):1082–6.
[11] Liu Y, Du H, Liu L, Leng J. Shape memory polymers and their composites in aerospace applications: a review. Smart Mater Struct 2014;23(2):23001–23.
[12] Sokolowski W, Metcalfe A, Hayashi S, Yahia L, Raymond J. Medical applications of shape memory polymers. Biomed Mater 2007;2(1):S23–7.
[13] Mano JF. Stimuli-responsive polymeric systems for biomedical applications. Adv Eng Mater. 2008;10(6):515–27.
[14] Lendlein A, Langer R, Bogart S. Biodegradable, Elastic shape-memory polymers for potential biomedical applications. Polymer 2016;296(5573):1673–6.
[15] Wang K, Strandman S, Zhu XX. A mini review: shape memory polymers for biomedical applications. Front Chem Sci Eng. 2017;11(2):1–11.
[16] Chen M, Tsai H, Chang Y, Lai W, Mi F. Rapidly self-expandable polymeric stents with a shape-memory property. Biomacromolecules 2007;2774–80.
[17] Xue L, Dai S LZ. Synthesis and characterization of three-arm poly(3-caprolactone)- based poly (ester urethanes) with shape-memory effect at body temperature. Biomaterials. 2009;42:964–72.
[18] Liu C, Qin H, Mather PT. Review of progress in shape-memory polymers. J Mater Chem 2007;17(16):1543.
[19] Lu XL, Cai W, Gao Z, Tang WJ. Shape memory effects of poly(L-lactide) and its copolymer with poly(ε-caprolactone). Polym Bull. 2007;58(2):381–91.
[20] Luo X, Mather PT. Preparation and characterization of shape memory elastomeric composites. Macromolecules. 2009;42(19):7251–3.
[21] Chen S, Hu J, Yuen C wah, Chan L. Novel moisture-sensitive shape memory polyurethanes containing pyridine moieties. Polymer 2009;50(19):4424–8.
[22] Wu L, Jin C, Sun X. Synthesis, properties, and light-induced shape memory effect of multiblock polyester urethanes containing biodegradable segments and pendant cinnamamide groups. Biomacromolecules 2011;12(1):235–41.
[23] Lee H-F, Yu HH. Study of electroactive shape memory polyurethane–carbon nanotube hybrids. Soft Matter 2011;7(8):3801.
[24] Schmidt AM. Electromagnetic activation of shape memory polymer networks containing magnetic nanoparticles. Macromol Rapid Commun 2006;27(14):1168–72.
[25] Weng S, Xia Z, Chen J, Gong L. Shape memory properties of polycaprolactone-based polyurethanes prepared by reactive extrusion. J Appl Polym Sci 2013;127(1):748–59.
[26] Sun L, Huang WM, Wang CC, Zhao Y, Ding Z, Purnawali H. Optimization of the shape memory effect in shape memory polymers. J Polym Sci Part A Polym Chem 2011;49(16):3574–81.
[27] Zhang J, Wu G, Huang C, Niu Y, Chen C, Chen H, et al. Unique multifunctional thermally-induced shape memory poly(p-dioxanone)-poly(tetramethylene oxide)glycol multiblock copolymers based on the synergistic effect of two segments. J Phys Chem C 2012;116(9):5835–45.
[28] Behl M, Razzaq MY, Lendlein A. Multifunctional shape-memory polymers. Adv Mater 2010;22(31):3388–410.
[29] Santiago D, Fabregat-Sanjuan A, Ferrando F, De la Flor S. Hyperbranched-modified epoxy thermosets: enhancement of thermomechanical and shape-memory performances. J Appl Polym Sci 2017;134(12):1–11.

[30] Santiago D, Fabregat-Sanjuan A, Ferrando F, De La Flor S. Recovery stress and work output in hyperbranched poly(ethyleneimine)-modified shape-memory epoxy polymers. J Polym Sci Part B Polym Phys 2016;54(10):1002–13.

[31] Santiago D, Fabregat-Sanjuan A, Ferrando F, De la Flor S. Improving of mechanical and shape-memory properties in hyperbranched epoxy shape-memory polymers. Shape Mem Superelasticity 2016;2(3):239–46.

[32] Jeong HJ, Kim BK. Shape memory hyperbranched polyurethanes via thiol-ene click chemistry. React Funct Polym 2017;116:92–100.

[33] Gao C, Yan D. Hyperbranched polymers: from synthesis to applications. Prog Polym Sci 2004;29(3):183–275.

[34] Thakur S, Karak N. Bio-based tough hyperbranched polyurethane-graphene oxide nanocomposites as advanced shape memory materials. RSC Advances 2013;3:9476–82.

[35] Voit B, Voit BI, Lederer A. Hyperbranched and highly branched polymer architectures – synthetic strategies and major characterization aspects. Chem Rev 2009;109:5924–73.

[36] Li H, Sivasankarapillai G, McDonald AG. Highly biobased thermally-stimulated shape memory copolymeric elastomers derived from lignin and glycerol-adipic acid based hyperbranched prepolymer. Ind Crops Prod 2015;67:143–54.

[37] Santiago D, Fernández-Francos X, Ferrando F, De La Flor S. Shape-memory effect in hyperbranched poly(ethyleneimine)-modified epoxy thermosets. J Polym Sci Part B Polym Phys 2015;53(13):924–33.

[38] Sivakumar C, Nasar AS. Poly(ε-caprolactone)-based hyperbranched polyurethanes prepared via A2 + B3 approach and its shape-memory behavior. Eur Polym J 2009;45(8):2329–37.

[39] Bayan R, Karak N. Renewable resource modified polyol derived aliphatic hyperbranched polyurethane as a biodegradable and UV-resistant smart material. Polym Int 2017;66(6):839–50.

[40] Kalita H, Karak N. Mesua ferrea L. seed oil-based hyperbranched shape memory polyurethanes: effect of multifunctional component. Polym Eng Sci 2012;52:2454–61.

[41] Duarah R, Singh YP, Mandal BB, Karak N, Deka H, Karak N, et al. Sustainable starch modified polyol based tough, biocompatible, hyperbranched polyurethane with a shape memory attribute. New J Chem 2016;40(6):5152–63.

[42] Liu G, Ding X, Cao Y, Zheng Z, Peng Y. Novel shape-memory polymer with two transition temperatures. Macromol Rapid Commun 2005;26(8):649–52.

[43] Shidhaye S, Surve C, Dhone A, Budhkar T. Interpenetrating polymer network-an overview. Int J Res Rev Pharm Appl Sci 2007;2(4):637–50.

[44] Dave VJ, Patel HS. Synthesis and characterization of interpenetrating polymer networks from transesterified castor oil based polyurethane and polystyrene. J Saudi Chem Soc 2017;21(1):18–24.

[45] Shivashankar M, Mandal BK. A review on interpenetrating polymer network. Int J Pharm Pharm Sci 2012;4:1–7.

[46] Sperling LH. Interpenetrating polymer networks: an overview. Adv Chem 1994;3–38.

[47] Archana D, Manisha G, Divya G. A review on interpenetrating polymer network (IPN). J Pharm Pharm Sci 2015;4(12):389–99.

[48] Ratna D, Karger-Kocsis J. Shape memory polymer system of semi-interpenetrating network structure composed of crosslinked poly (methyl methacrylate) and poly (ethylene oxide). Polymer 2011;52(4):1063–70.

[49] Zhang S, Feng Y, Zhang L, Xu Y. Novel interpenetrating networks with shape-memory properties. J Polym Sci Part A: Polym Chem 2007;45:768–75.

[50] Liu G, He W, Peng Y, Xia H. Shape-memory behavior of poly (methyl methacrylate-co –N-vinyl-2-pyrrolidone) / poly (ethylene glycol) semi-interpenetrating polymer networks based on hydrogen bonding. J Polym Res 2011;18(6):2109–17.

[51] Li J, Liu T, Xia S, Pan Y, Zheng Z, Ding X, et al. A versatile approach to achieve quintuple-shape memory effect by semi-interpenetrating polymer networks containing broadened glass transition and crystalline segments. J Mater Chem 2011;21(33):12213–7.
[52] Cha KJ, Lih E, Choi J, Joung YK, Ahn DJ, Han DK. Shape-memory effect by specific biodegradable polymer blending for biomedical applications. Macromol Biosci 2014;14(5):667–78.
[53] Jing X, Mi HY, Huang HX, Turng LS. Shape memory thermoplastic polyurethane (TPU)/ poly(ε-caprolactone) (PCL) blends as self-knotting sutures. J Mech Behav Biomed Mater 2016;64:94–103.
[54] Ahmed N, Kausar A, Muhammad B. Shape memory properties of electrically conductive multi-walled carbon nanotube-filled polyurethane/modified polystyrene blends. J Plast Film Sheeting 2016;32(3):272–92.
[55] Sattar R, Kausar A, Siddiq M. Thermal, mechanical and electrical studies of novel shape memory polyurethane/polyaniline blends. Chin J Polym Sci 2015;33(9):1313–24.
[56] Cavicchi KA. Shape memory polymers from blends of elastomers and small molecule additives. Macromol Symp 2015;358(1):194–201.
[57] Kashif M, Kim S-J, Chang Y-W. Shape memory polymer blends of syndiotactic 1,2-polybutadiene and trans-polyoctenamer. Polym Bull 2017;74(7):2535–44.
[58] Samuel C, Barrau S, Lefebvre JM, Raquez JM, Dubois P. Designing multiple-shape memory polymers with miscible polymer blends: evidence and origins of a triple-shape memory effect for miscible PLLA/PMMA blends. Macromolecules 2014;47(19):6791–803.
[59] Meng Q, Hu J. A review of shape memory polymer composites and blends. Compos Part A Appl Sci Manuf 2009;40(11):1661–72.
[60] Jeong HM, Song JH, Lee SY, Kim BK. Miscibility and shape memory property of poly (vinyl chloride)/ thermoplastic polyurethane blends. J Mater Sci 2001;6(36):5457–63.
[61] Jeong HM, Ahn BK, Kim BK. Miscibility and shape memory effect of thermoplastic polyurethane blends with phenoxy resin. Eur Polym J 2001;37(11):2245–52.
[62] Erden N, Jana SC. Synthesis and characterization of shape-memory polyurethane-polybenzoxazine compounds. Macromol Chem Phys 2013;214(11):1225–37.
[63] Mather PT. Blends of amorphous and semicrystalline polymers having shape memory properties. Patent US 7795350 B2 2010.
[64] You J, Dong W, Zhao L, Cao X, Qiu J, Sheng W, et al. Crystal orientation behavior and shape-memory performance of poly (vinylidene fluoride)/acrylic copolymer blends. J Phys Chem B 2012;116:1256–64.
[65] Behl M, Ridder U, Feng Y, Kelch S, Lendlein A. Shape-memory capability of binary multiblock copolymer blends with hard and switching domains provided by different components. Soft Matter 2009;5(3):676.
[66] Li SC, Lu LN, Zeng W. Thermostimulative shape-memory effect of reactive compatibilized high-density polyethylene/poly(ethylene terephthalate) blends by an ethylene–butyl acrylate–glycidyl methacrylate terpolymer. J Appl Polym Sci 2009;112(6):3341–46.
[67] Chatterjee T, Dey P, Nando GB, Naskar K. Thermo-responsive shape memory polymer blends based on alpha olefin and ethylene propylene diene rubber. Polymer 2015;78:180–92.
[68] Zhang H, Wang H, Zhong W, Du Q. A novel type of shape memory polymer blend and the shape memory mechanism. Polymer 2009;50(6):1596–601.
[69] Weiss RA, Izzo E, Mandelbaum S. New design of shape memory polymers: mixtures of an elastomeric ionomer and low molar mass fatty acids and their salts. Macromolecules 2008;41(9):2978–80.
[70] Lligadas G, Ronda JC, Gali M, Cdiz V. Renewable polymeric materials from vegetable oils: a perspective. Mater Today 2013;16(9):337–43.
[71] Adekunle KF. A review of vegetable oil-based polymers : synthesis and applications. Open J Polym Chem 2015;5:34–40.

[72] Nawrath C, Poirier Y, Somerville C. Plant polymers for biodegradable plastics: cellulose, starch and polyhydroxyalkanoates. Mol Breed 1995;1(2):105–22.
[73] Rashmi BJ, Loux C, Prashantha K. Bio-based thermoplastic polyurethane and polyamide 11 bioalloys with excellent shape memory behavior. J Appl Polym Sci 2017;134(20):1–10.
[74] Del Río E, Lligadas G, Ronda JC, Galià M, Cádiz V, Meier MAR. Shape memory polyurethanes from renewable polyols obtained by ATMET polymerization of glyceryl triundec-10-enoate and 10-undecenol. Macromol Chem Phys 2011;212(13):1392–9.
[75] Tsujimoto T, Ohta E, Uyama H. Plant oil-based shape memory polymer using acrylic monolith. Express Polym Lett 2015;9(9):757–63.
[76] Petrović ZS, Milić J, Zhang F, Ilavsky J. Fast-responding bio-based shape memory thermoplastic polyurethanes. Polymer 2017;121:26–37.
[77] Tsujimoto T, Takayama T, Uyama H. Biodegradable shape memory polymeric material from epoxidized soybean oil and polycaprolactone. Polymer 2015;7(10):2165–74.
[78] Saralegi A, Johan Foster E, Weder C, Eceiza A, Corcuera MA. Thermoplastic shape-memory polyurethanes based on natural oils. Smart Mater Struct 2014;23(2):25033.
[79] Yuan D, Chen Z, Xu C, Chen K, Chen Y. Fully biobased shape memory material based on novel cocontinuous structure in poly(lactic acid)/natural rubber TPVs fabricated via peroxide-induced dynamic vulcanization and in situ interfacial compatibilization. ACS Sustain Chem Eng 2015;3(11):2856–65.
[80] Desroches M, Escouvois M, Auvergne R, Caillol S, Boutevin B. From vegetable oils to polyurethanes: synthetic routes to polyols and main industrial products. Polym Rev 2012;52(1):38–79.
[81] Xia Y, Larock RC. Vegetable oil-based polymeric materials: synthesis, properties, and applications. Green Chem 2010;12(11):1893.
[82] Petrovic Z. Polyurethanes from vegetable oils. Polym Rev 2008;48(1):109–55.
[83] Li F, Larock RC. New soybean oil-styrene-divinylbenzene thermosetting copolymers. V. Shape memory effect. J Appl Polym Sci 2002;84(8):1533–43.
[84] Guo B, Chen Y, Lei Y, Zhang L, Zhou WY, Rabie ABM. Biobased poly (propylene sebacate) as shape memory polymer with tunable switching temperature for potential biomedical applications. Biomacromolecules 2011;12(4):1312–21.
[85] Goerz O, Ritter H. Polymers with shape memory effect from renewable resources: crosslinking of polyesters based on isosorbide, itaconic acid and succinic acid. Polym Int 2013;62(5):709–12.
[86] Cai W, Liu L. Shape-memory effect of poly (glycerol-sebacate) elastomer. Mater Lett 2008;62(14):2175–7.
[87] Zhang K, Geissler A, Standhardt M, Mehlhase S, Gallei M, Chen L, et al. Moisture-responsive films of cellulose stearoyl esters showing reversible shape transitions. Sci Rep 2015;5:11011.
[88] Kalita H, Karak N. Biobased hyperbranched shape-memory polyurethanes: effect of different vegetable oils. J Appl Polym Sci 2014;131(1):1–8.
[89] Jimenez GA, Jana SC. Composites of carbon nanofibers and thermoplastic polyurethanes with shape-memory properties prepared by chaotic mixing. Polym Eng Sci 2009;49:2020–30.
[90] Keledi G, Hári J, Pukánszky B. Polymer nanocomposites: structure, interaction, and functionality. Nanoscale 2012;4(6):1919.
[91] Jordan J, Jacob KI, Tannenbaum R, Sharaf MA, Jasiuk I. Experimental trends in polymer nanocomposites – a review. Mater Sci Eng A 2005;393(1–2):1–11.
[92] Mittal V. Optimization of polymer nanocomposite properties. New Jersey: Wiley, 2010.
[93] Liu T, Zhou T, Yao Y, Zhang F, Liu L, Liu Y, et al. Stimulus methods of multi-functional shape memory polymer nanocomposites: a review. Compos Part A Appl Sci Manuf 2017;100:20–30.
[94] Wang W, Liu Y, Leng J. Recent developments in shape memory polymer nanocomposites: actuation methods and mechanisms. Coord Chem Rev 2016;320–321:38–52.

[95] Barwood MJ, Breen C, Clegg F, Hammond CL. The effect of organoclay addition on the properties of an acrylate based, thermally activated shape memory polymer. Appl Clay Sci 2014;102:41–50.
[96] Xu B, Fu YQ, Huang WM, Pei YT, Chen ZG, Hosson JD, et al. Thermal-mechanical properties of polyurethane-clay shape memory polymer nanocomposites. Polymer 2010;2(2):31–9.
[97] Pavlidou S, Papaspyrides CD. A review on polymer-layered silicate nanocomposites. Prog Polym Sci 2008;33(12):1119–98.
[98] Cao F, Jana SC. Nanoclay-tethered shape memory polyurethane nanocomposites. Polymer 2007;48(13):3790–800.
[99] Pan GH, Huang WM, Ng ZC, Liu N, Phee SJ. The glass transition temperature of polyurethane shape memory polymer reinforced with treated/non-treated attapulgite (palygorskite) clay in dry and wet conditions. Smart Mater Struct 2008;17(4):45007–13.
[100] Kim MS, Jun JK, Jeong HM. Shape memory and physical properties of poly(ethyl methacrylate)/Na-MMT nanocomposites prepared by macroazoinitiator intercalated in Na-MMT. Compos Sci Technol 2008;68(7–8):1919–26.
[101] Coativy G, Gautier N, Pontoire B, Buléon A, Lourdin D, Leroy E. Shape memory starch-clay bionanocomposites. Carbohydr Polym 2015;116:307–13.
[102] Subramoney BS. Novel nanocarbons structure, properties, and potential applications. Adv Mater 1998;(15):1157–71.
[103] Yu M-F, Files BS, Arepalli S, Ruoff RS. Tensile loading of ropes of single wall carbon nanotubes and their mechanical properties. Phys Rev Lett 2000;84(24):5552–5.
[104] Yu M. Strength and breaking mechanism of multiwalled carbon nanotubes under tensile load. Science 2000;287(5453):637–40.
[105] Liew KM, Lei ZX, Zhang LW. Mechanical analysis of functionally graded carbon nanotube reinforced composites: a review. Compos Struct 2015;120:90–7.
[106] Liu Y, Zhao J, Zhao L, Li W, Zhang H, Yu X, et al. High performance shape memory epoxy/carbon nanotube nanocomposites. ACS Appl Mater Interfaces 2016;8(1):311–20.
[107] Coleman JN, Khan U, Blau WJ, Gun'ko YK. Small but strong: a review of the mechanical properties of carbon nanotube-polymer composites. Carbon 2006;44(9):1624–52.
[108] Li H, Zhong J, Meng J, Xian G. The reinforcement efficiency of carbon nanotubes/shape memory polymer nanocomposites. Compos Part B Eng 2013;44(1):508–16.
[109] Gu S, Yan B, Liu L, Ren J. Carbon nanotube-polyurethane shape memory nanocomposites with low trigger temperature. Eur Polym J 2013;49(12):3867–77.
[110] Kalita H, Karak, N. Bio-based hyperbranched polyurethane/multi-walled carbon nanotube nanocomposites as shape memory materials. Polym Compos 2014;35(4):636–43.
[111] Kalita H, Karak N. Hyperbranched polyurethane/Fe_3O_4 nanoparticles decorated multiwalled carbon nanotube thermosetting nanocomposites as microwave actuated shape memory materials. J Mater Res 2013;28(16):2132–41.
[112] Jin Yoo H, Chae Jung Y, Gopal Sahoo N, Whan Cho J. Polyurethane-carbon nanotube nanocomposites prepared by in-situ polymerization with electroactive shape memory. J Macromol Sci Part B 2006;45(4):441–51.
[113] Lu H, Liu Y, Gou J, Leng J, Du S. Electroactive shape-memory polymer nanocomposites incorporating carbon nanofiber paper. Int J Smart Nano Mater 2010;1:2–12.
[114] Lu H, Zhao D, Tang Y, Gou J, Leng J, Du S. Shape memory polymer carbon nanocomposites; based on carbon nanopaper. Earth Sp 2010;3665–78.
[115] Xie F, Huang L, Leng J, Liu Y. Thermoset shape memory polymers and their composites. J Intell Mater Syst Struct 2016;27(18):2433–55.
[116] De Jong KP, Geus JW, Jong K De. Carbon nanofibers: catalytic synthesis and applications. Catal Rev 2000;42:481–510.

[117] Feng L, Xie N, Zhong J. Carbon nanofibers and their composites: a review of synthesizing, properties and applications. Materials 2014;7(5):3919–45.

[118] Luo X, Mather PT. Conductive shape memory nanocomposites for high speed electrical actuation. Soft Matter 2010;6(10):2146.

[119] Tang Z, Sun D, Yang D, Guo B, Zhang L, Jia D. Vapor grown carbon nanofiber reinforced bio-based polyester for electroactive shape memory performance. Compos Sci Technol 2013;75:15–21.

[120] Balandin AA, Ghosh S, Bao W, Calizo I, Teweldebrhan D, Miao F, et al. Superior thermal conductivity of single-layer graphene. Nano Lett 2008;8(3):902–7.

[121] Stoller MD, Park S, Zhu Y, An J, Ruoff RS. Graphene-based ultracapacitors. Nano Lett 2008;8(10):3498–502.

[122] Singh V, Joung D, Zhai L, Das S, Khondaker SI, Seal S. Graphene based materials: past, present and future. Prog Mater Sci 2011;56(8):1178–271.

[123] Lee C, Wei X, Kysar JW, Hone J. Measurement of the elastic properties and intrinsic strength of monolayer graphene. Sci Rep 2008;321(5887):385–8.

[124] Shams SS, Zhang R, Zhu J. Graphene synthesis: a Review. Mater Sci Pol 2015;33(3):566–78.

[125] Dhand V, Rhee KY, Kim HJ, Jung DH. A comprehensive review of graphene nanocomposites: research status and trends. J Nanomater 2013;2013:1–14.

[126] Yoonessi M, Shi Y, Scheiman D, Colon M, Tigelaar D, Weiss RA, et al. Graphene polyimide nanocomposites; high-temperature shape memory effects. ACS Nano 2012;6(9):7644–55.

[127] Aïssa B, Memon NK, Ali A, Khraisheh MK. Recent progress in the growth and applications of graphene as a smart material: a review. Front Mater 2015;2:1–19.

[128] Zhang Z, Dou J, He J, Xiao C, Shen L, Yang J, et al. Electrically/infrared actuated shape memory composites based on a bio-based polyester blend and graphene nanoplatelets and their excellent self-driven ability. J Mater Chem C 2017;5(17):4145–58.

[129] Chan BQY, Low ZWK, Heng SJW, Chan SY, Owh C, Loh XJ. Recent advances in shape memory soft materials for biomedical applications. ACS Appl Mater Interfaces 2016;8(16):10070–87.

[130] Jung YC, Kim JH, Hayashi T, Kim YA, Endo M, Terrones M, et al. Fabrication of transparent, tough, and conductive shape-memory polyurethane films by incorporating a small amount of high-quality graphene. Macromol Rapid Commun 2012;33(8):628–34.

[131] Lashgari S, Karrabi M, Ghasemi I, Azizi H, Messori M, Paderni K. Shape memory nanocomposite of poly(L-lactic acid)/graphene nanoplatelets triggered by infrared light and thermal heating. Express Polym Lett 2016;10(4):349–59.

[132] Tang Z, Kang H, Wei Q, Guo B, Zhang L, Jia D. Incorporation of graphene into polyester/carbon nanofibers composites for better multi-stimuli responsive shape memory performances. Carbon 2013;64:487–98.

[133] Han S, Chun BC. Preparation of polyurethane nanocomposites via covalent incorporation of functionalized graphene and its shape memory effect. Compos Part A Appl Sci Manuf 2014;58:65–72.

[134] Liu Y, Gall K, Dunn ML, McCluskey P. Thermomechanics of shape memory polymer nanocomposites. Mech Mater 2004;36(10):929–40.

[135] Gunes IS, Cao F, Jana SC. Evaluation of nanoparticulate fillers for development of shape memory polyurethane nanocomposites. Polymer 2008;49(9):2223–34.

[136] Likitaporn C, Mora P, Tiptipakorn S, Rimdusit S. Recovery stress enhancement in shape memory composites from silicon carbide whisker–filled benzoxazine-epoxy polymer alloy. J Intell Mater Syst Struct 2017;29(3):388–96.

[137] Du H, Song Z, Wang J, Liang Z, Shen Y, You F. Microwave-induced shape-memory effect of silicon carbide/poly(vinyl alcohol) composite. Sensors Actuators, A Phys 2015;228:1–8.

[138] Li J, Shirai T, Fuji M. Silicon carbide and its nanostructure. Adv Ceram Res Cent Ann Rep 2014;3:5–10.
[139] Gall K, Dunn ML, Liu Y, Stefanic G, Balzar D. Internal stress storage in shape memory polymer nanocomposites. Appl Phys Lett 2004;85(2):290–2.
[140] Gall K, Dunn ML, Liu Y, Finch D, Lake M, Munshi NA. Shape memory polymer nanocomposites. Acta Mater 2002;50(20):5115–26.
[141] Lee A, Lichtenhan J. Viscoelastic responses of polyhedral oligosilsesquioxane reinforced epoxy systems. Macromolecules 1998;31(15):4970–4.
[142] Romo-Uribe A, Mather PT, Haddad TS, Lichtenhan JD. Viscoelastic and morphological behavior of hybrid styryl-based polyhedral oligomeric silsesquioxane (POSS) copolymers. J Polym Sci Part B Polym Phys 1998;36(11):1857–72.
[143] Kim GM, Qin H, Fang X, Sun FC, Mather PT. Hybrid epoxy-based thermosets based on polyhedral oligosilsesquioxane: cure behavior and toughening mechanisms. J Polym Sci Part B Polym Phys 2003;41(24):3299–313.
[144] Jeon HG, Mather PT, Haddad TS. Shape memory and nanostructure in poly (norbornyl-POSS) copolymers. Polym Int 2000;457:453–7.
[145] Gu SY, Jin SP, Liu LL. Polyurethane/polyhedral oligomeric silsesquioxane shape memory nanocomposites with low trigger temperature and quick response. J Polym Res 2015;22(7):142–8.
[146] Haxton KJ, Morris RE. Polyhedral oligomeric silsesquioxane dendrimers. New York: Springer, 2009:121–39.
[147] Pielichowski K, Njuguna J, Janowski B, Pielichowski J. Polyhedral oligomeric silsesquioxanes (POSS)-containing nanohybrid polymers. Adv Polym Sci 2006;201(1):225–96.
[148] Knight PT, Lee KM, Chung T, Mather PT. PLGA-POSS end-linked networks with tailored degradation and shape memory behavior. Macromolecules 2009;42(17):6596–604.
[149] Knight PT, Lee KM, Qin H, Mather PT. Biodegradable thermoplastic polyurethanes incorporating polyhedral oligosilsesquioxane. Biomacromolecules 2008;9(9):2458–67.
[150] Rattinger EH, Ishida K, Uribe AR, Mather, PT. Thermally modulated nanostructure of poly(ε-caprolactone)-POSS multiblock thermoplastic polyurethanes. Polymer 2013;54:3350–62.
[151] Fu BX, Hsiao BS, Pagola S, Stephens P, White H, Rafailovich M, et al. Structural development during deformation of polyurethane containing polyhedral oligomeric silsesquioxanes (POSS) molecules. Polymer 2001;42:599–611.
[152] Mya KY, Gose HB, Pretsch T, Bothe M, He C. Star-shaped POSS-polycaprolactone polyurethanes and their shape memory performance. J Mater Chem 2011;21(13):4827–36.

2 Classification of shape memory polymers

2.1 Introduction

Shape memory polymers (SMPs) can be classified based on the nature of their structures such as amorphous or crystalline and covalently cross-linked or physically cross-linked network (Figure 2.1) [1, 2]. The different SMPs show dissimilar shape memory effect (SME) because of their different nature of structures. SME of SMP depends on the length and functionality of the cross-links, nature of switching segment, crystallinity, crystalline structure and network structure. T_{trans} of SMP can be either T_g or T_m based on their structure. SMP having T_g far below room temperature, T_m, is used as T_{trans}. Normally T_g-based SMPs show slower SME as compared to T_m-based SMP because of the broad glass transition. This classification of SMP plays an important role in interpreting the underlying mechanisms of the SME, and subsequently in understanding the design of new SMP. SMP can also be categorized based on their external stimulus. SMPs possess at least two phases, one is permanent network which is responsible for retaining the original shape and the second one is reversible network which is responsible for fixing the temporary shape. Based on the nature and chemical structure of reversible phase, the external stimuli can be chosen. This categorization helps selection of SMP for their suitable applications. The classification of SMP is discussed below.

2.2 Based on nature of structure

2.2.1 Covalently cross-linked glassy SMP

This type of SMP is covalently cross-linked glassy polymer below T_g and rubbery elastic above T_g. Covalent cross-links can be obtained during synthesis or by post-processing methods whereby at least one monomer should have more than two functional groups. The nature of permanent cross-linking leads to excellent shape fixity and shape recovery of this SMP, which can be adjusted through the extent of cross-linking. The excellent shape fixity and recovery of SMP are caused by the high modulus below T_g and excellent rubbery elasticity above T_g. However, once finally processed (casting or molding), these materials are difficult to reshape again, as the primary shape is covalently fixed. Shape memory properties of this SMP depend on the length and functionality of the cross-links, nature of switching segment and network structure. The mechanism of SME of this type of SMP is shown in Figure 2.2. Typical examples are thermosetting styrene–butadiene copolymer, polyethylene, thermosetting polyurethanes (PUs), epoxy and so on (Table 2.1). An SMP based on poly(vinyl alcohol) (PVA) chemically cross-linking with glutaraldehyde (GA) has been reported [3]. PVA is a typical semicrystalline polymer,

https://doi.org/10.1515/9783110570175-002

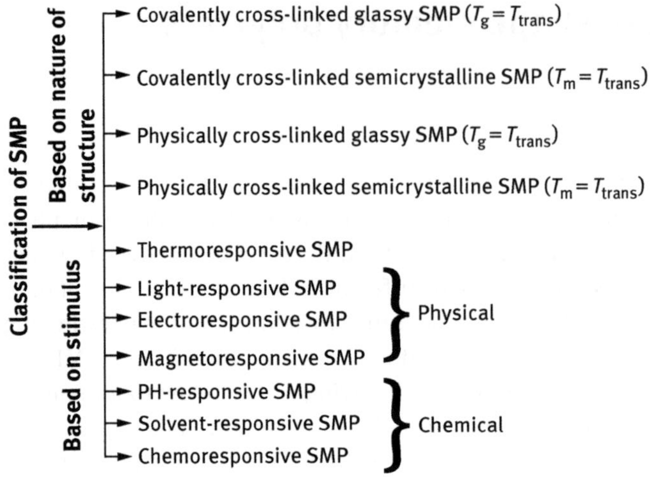

Figure 2.1: A presentation of classification of SMP.

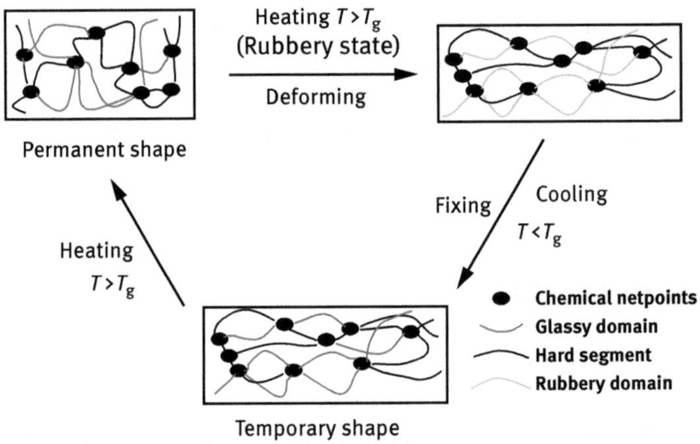

Figure 2.2: The mechanism of SME of covalently cross-linked glassy SMP.

owing to the hydrogen bonding between hydroxyl groups. However, crystallinity was absent when the GA concentration was more than 3 wt%. This can be attributed to the high cross-link density due to the presence of more GA content. They have exhibited excellent shape recovery ratio even after several test cycles. The chemical and physical networks serve as the fixed phase, while the amorphous region acts as the reversible phase. Unsaturated polyesters chemically cross-linking with methacrylate SMPs have been developed [4]. α-Methylene-γ-butyrolactone was synthesized from γ-butyrolactone. α-Methylene-γ-butyrolactone was copolymerized with ε-caprolactone (ε-CL) to synthesize unsaturated polyester poly(α-methylene-γ-butyrolactone-co-ε-caprolactone) via ring-opening reaction. This unsaturated polyester was cross-linked with methacrylates by radical reactions. T_g of these materials can be tuned in a range of –26

Table 2.1: Examples of covalently cross-linked glassy and semicrystalline SMP.

SMP	Hard segment	Soft segment	T_{trans} (°C)
Covalently cross-linked glassy SMP			
Poly(vinyl alcohol) glutaraldehyde	Cross-link	Poly(vinyl alcohol)	60–65
Polyesters-methacrylates	Cross-link	Methacrylates	−26 to 29
Poly(ethylene terephthalate)-co-poly(ethylene glycol)- glycerin	Poly(ethylene terephthalate))	Poly(ethylene glycol)	11–24
Epoxy–Jeffamines	Epoxy	Jeffamines	31–93
Copolyester–polyurethane	1,6-Diisocyanato-2,2,4-trimethylhexane	Oligo[(rac-lactide)-co-glycolide]	36–60
Methyl methacrylate-co-PEGDMA	Cross-link	PEGDMA	56–92
Starch–glycerol	Cross-link	Starch	50
PU-PB	MDI/PB/1,4-butanediol	Poly(tetramethylene) glycol	70–110
Covalently cross-linked semicrystalline SMP			
Dicumyl peroxide- polycyclooctene	Cross-link	Polycyclooctene	30–60
PU-epoxy	TDI/1,4-butanediol	PCL	30–40
Poly(ethylene)-co-1-octene)	Cross-link	Poly(ethylene)-co-1-octene	60–100
Alkoxysilane-poly (ε-caprolactone)	Cross-link	PCL	40–60
Oligo[(ε-hydroxycaproate)-co-glycolate]dimethacrylates	Cross-link	Oligo[(ε-hydroxycaproate)-co-glycolate]dimethacrylates	18–52
Natural rubber	Cross-link	Polyisoprene	0–45
LDPE	Cross-link	LDPE	60–100

PEGDMA, poly(ethylene glycol) dimethacrylate; PB, polybenzoxazine; MDI, methylene diphenyl diisocyanate; TDI, toluene diisocyanate; PU, polyurethane; PCL, polycaprolactone; LDPE, low-density polyethylene.

to 29 °C. The materials exhibited excellent shape memory behaviors. Shape memory poly(ethylene terephthalate)) and poly(ethylene glycol) (PEG) copolymers cross-linking with glycerin, sorbitol or maleic anhydride have been developed [5]. The copolymer with 2.5 mol% of glycerin exhibited the highest shape recovery behavior. The following suggested characteristics are responsible for high shape recovery behavior of the cross-linked copolymer: (1) the molecular interactions among hard segments (poly(ethylene terephthalate)) [PET] part), for example, H-bonding, dipole–dipole interaction of carbonyl groups of PET and induced dipole–dipole interaction between aromatic rings; (2) the soft segment (PEG part) efficiently and entirely absorbs the applied load as a result, and the relative position of the hard segment or the original shape of the copolymer can be safely preserved under mechanical deformation; (3) glycerin is more effective for SME than sorbitol as earlier is a shorter cross-linking agent, although both are linking with the hard segments, and the short and rigid structure of glycerin is more favored for shape retention of the hard segment. Cross-link density has a strong influence on the shape memory properties of the copolymers. T_g can be controlled by adjustment of the composition of the copolymer and cross-linking agent. Shape memory epoxy based on Jeffamines curing oligo (bisphenol A) diglycidyl ether has been developed [6]. The materials exhibited excellent shape fixity and shape recovery behaviors to almost 100%. The main structural factors for epoxy to exhibit the shape memory performances are to reduce the cross-link density, a high-chain flexibility and molecular mobility. The transition temperature, i.e., T_g can be tuned in the range of 31–93 °C. The higher cross-linking density of these materials exhibited the faster shape recovery. Lendlein et al. developed copolyester-PU thermoset SMPs, which are prepared from star-shaped oligo[(rac-lactide)-co-glycolide] soft segments and low-molecular-weight diisocyanate [7]. The materials exhibited shape fixity and shape recovery was more than 90%. The shape memory properties found to vary slightly between the first and the second cycles of test because of the segment chain orientation and relaxation effects. The T_g can be modified in the range of 36–60 °C. Methyl methacrylate and PEG dimethacrylate (PEGDMA) copolymer thermoset SMPs have been reported [8]. The T_g and rubbery modulus of the SMP can be controlled by adjusting the amount and molecular weight of PEGDMA cross-linking monomer. T_g can be adjusted in the range of 56–92 °C.

2.2.2 Covalently cross-linked semicrystalline SMP

It is similar to the first category of SMP, where the permanent shapes are established through the cross-linking and cannot be reshaped once setting. However, the temporary shape can be controlled through deformation above T_m of the crystallization region and successive cooling below the T_m. This is because of the fact that both T_g and T_m can be used as a T_{trans} for this kind of SMP, though most often T_m is used, as stated in Chapter 1. The shape fixity of this kind of SMP can be improved by choosing a relatively lower temperature compared to T_{trans} and that low temperature should be such so that it allows a high degree of crystallization. The shape recovery speed is generally faster for this kind of SMP as T_m is a first-order transition and a sharp transition. The shape memory behaviors of this SMP are influenced by the functionality of cross-link, crystallinity, crystalline structure and network structure. The mechanism of SME of this type of SMP is shown in Figure 2.3. This class of SMP includes bulk polymers, like semicrystalline polymers, liquid crystal elastomers and hydrogels with phase-separated crystalline microstructure. Thermosetting semicrystalline polyisoprene, polycyclooctenes and polycaprolactone are examples of this class (Table 2.1). SMPs based on polycyclooctenes chemically cross-linking with dicumyl peroxide have been developed [9]. Polycyclooctene was synthesized through ring-opening metathesis polymerization of cyclooctene using the dihydroimidazolylidene-modified Grubb's catalyst. The neat polycyclooctene does not have SME due to the lack of a rubbery plateau above T_m. The cross-linked polycyclooctene exhibited shape memory behaviors, and the shape recovery was found to be increased with the increase in the cross-link density. According to the theory of rubbery elasticity, networks with higher cross-linking densities have higher driving forces to return to their lowest entropy

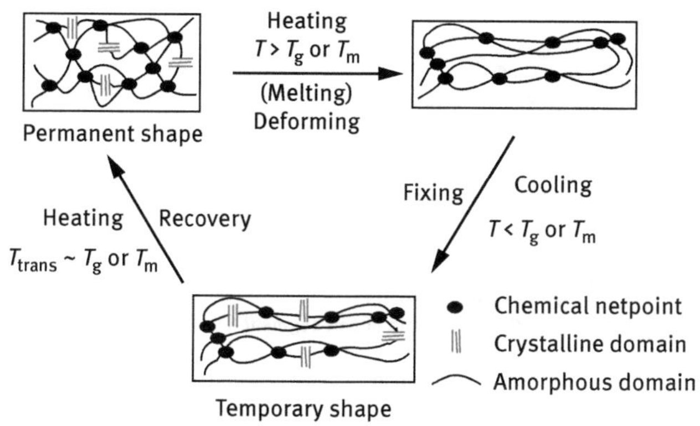

Figure 2.3: The mechanism of SME of covalently cross-linked semicrystalline SMP.

state. Hu et al. [10] reported the shape memory behaviors of chemically cross-linked PUs. PUs were chemically cross-linked with the excess amount of 4,4′-methylenebis(phenyl isocyanate) (MDI) and glycerin. The cross-linked PU performed better SME than the linear PU. Cross-linked poly(ethylene/1-octene) shape memory copolymers have been developed [11]. The switching temperature, i.e., T_m and crystallinity of the copolymer decreased with the increase of cross-link density. The switching temperature can be tailored in the range of 60–100 °C by varying the degree of branching and cross-link density. The shape recovery and shape fixity were increased with the increase of crystallinity and cross-link density. It was observed that crystallinity and the perfection of crystalline structure have strong influence to store the generated strain energy. A novel covalently cross-linked alkoxysilane-terminated poly(ε-CL) SMP has been reported [12]. Sol–gel process was used to prepare the covalently cross-linked SMP, where Si–O–Si domains behave as the net points essential for the shape memory behaviors. The shape memory behaviors of the polymer, especially recovery time, depend on the molecular weight of the poly(ε-CL) and cross-link density of the systems. Photo-cross-linked oligo[(ε-hydroxycaproate)-co-glycolate]dimethacrylate SMPs have been developed [13]. By varying the molecular weight and the content, glycolate of the switching segment, T_m, which correlates with T_{trans}, can be adjusted between 18 and 52 °C. In addition, shape memory behaviors of these SMPs were improved by formation of AB-type copolymer network with *n*-butyl acrylate. Study revealed that copolymer networks from oligo[(ε-hydroxycaproate)-co-glycolate]dimethacrylates with *Mn* of the cross-linked macromonomers of at least 6,900 g/mol and a 14 mol% of glycolate content exhibited excellent shape fixity with values more than 94%.

2.2.3 Physically cross-linked glassy SMP

In this SMP, the hard amorphous domains act as physical cross-links via van der Waals forces, polar–polar interactions, hydrogen bonding and so on and afford the

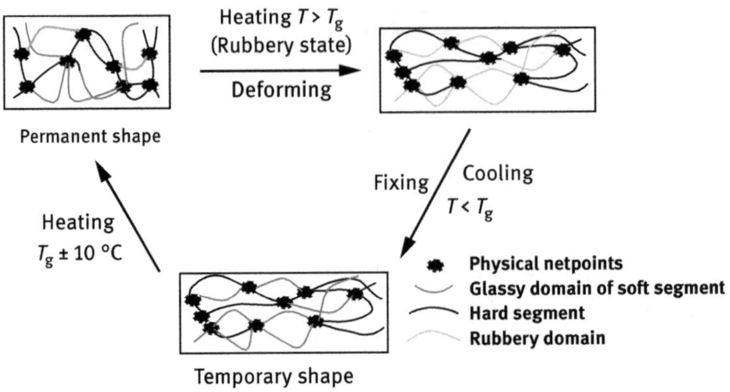

Figure 2.4: The mechanism of SME of physically cross-linked glassy SMP.

required elasticity for SME. These are mainly phase-separated linear block copolymers and multiblock copolymers or blends. T_{trans}, i.e., T_g of the soft segment is responsible for SME. When the temperature is above T_g of these discrete physical domains, the material will be deformed easily so that it can be processed and reshaped. For the temperature above T_g, it softens to a rubbery state and fixes a secondary shape on cooling below T_g. This type of SMP exhibits to some extent lower SME when compared with the other classes. This can be attributed to the loss of physical cross-link integrity due to the mechanical deformation. The shape memory behaviors of this SMP depend on the chain length, structure and arrangement of soft segment. The mechanism of SME of this type of SMP is shown in Figure 2.4. Most of the SMPs in this class are segmented amorphous shape memory PUs (Table 2.2). SMEs of physically cross-linked MDI and poly(tetramethylene glycol) (PTMG) soft segment-based shape memory PUs have been reported [14]. They used two types of soft segment, i.e., long and short with molecular weights PTMG-1800 and PTMG-1000, respectively. They studied the effect of arrangement of soft segments (random and block) on the SME of PUs. Random and block PUs showed better SME than copolymers with one kind of PTMG. The materials exhibited shape recovery and shape retention of more than 80%. Jeong et al. [15] reported the SME of polycaprolactone (PCL)-based thermoplastic polyurethane (TPU) blends with phenoxy resin. The PCL segment and phenoxy resin made single miscible domain with single T_g (T_g > RT) which comes between those of PCL segment and phenoxy resin. The blends showed the disappearance of crystallization and melting peaks of PCL segment which implied that the crystallization of PCL segment is severely hindered in the miscible blend. The blends showed good shape recovery behavior when actuated to a temperature of 70 °C. The miscible domain acts as a reversible phase, whereas phase-separated hard segment domain serves as a fixed structure memorizing the original shape. Poly(l-lactide-co-ε-caprolactone)-based biodegradable shape memory PUs have been developed. The material exhibited shape recovery

Table 2.2: Examples of physically cross-linked glassy and semicrystalline SMPs.

SMP	Hard segment	Soft segment	T_{trans} (°C)
Physically cross-linked glassy SMP			
PU–phenoxy resin	HDI/4,4′-dihydroxy biphenyl	PCL/phenoxy resin miscible domain	40–70
PU–PVC	HDI/4,4′-dihydroxy biphenyl	PCL/PVC miscible domain	40–50
Poly(ethylene adipate)–PU	MDI/1,4-butanediol	Poly(ethylene adipate)	10–50
Polystyrene–polybutadiene	Polystyrene	Polybutadiene	70–140
Poly(L-lactide-co-ε-caprolactone) -–PU	TDI/1,4-butanediol	Poly(L-lactide-co-ε-caprolactone)-	30–55
Poly(acrylic acid)-co-methyl methacrylate)/PEG	Cross-link	Poly(acrylic acid)/PEG blend	45–60
Polymethyl methacrylate-co-*N*-vinyl-2-pyrrolidone (VP)/PEG	Cross-link	PEG–PVP complex phase	80–100
PVDF/ACP	Netpoints	PVDF–ACP amorphous phase	45
Polytetramethylene oxide/ poly(acrylic acid-co-acrylonitrile)	Cross-link	Polytetramethylene oxide complex	60–100
Physically cross-linked semicrystalline SMP			
PEG–PU	MDI/1,4-butanediol	PEG	30–40
PCL–PU	TDI/1,4-butanediol	PCL	50–54
Three-arm PCL–PU	MDI/1,6-hexanediol	Three-arm PCL(ε-caprolactone + glycerol)	36–39
Poly(tetramethylene glycol)–PU	MDI/1,4-butanediol	Poly(tetramethylene glycol)	14–24
Poly(tetramethylene oxide) glycol–PU	MDI/1,4-butanediol	Poly(tetramethylene oxide) glycol	0–46
PEO–PET	PET	PEO	45–55
PE-nylon 6	Nylon 6	PE	100–140
PEEK	Crystalline domain	Amorphous domain	>180
EOC–EPDM	EOC	EPDM	≥60
PPDO-PCL	PPDO	PCL	>40

PPDO, poly(*p*-dioxanone); EOC, ethylene octene copolymer; EPDM, ethylene propylene diene rubber; PB, polybenzoxazine; MDI, methylene diphenyl diisocyanate; HDI, hexamethylene diisocyanate; TDI, toluene diisocyanate; PU, polyurethane; PCL, polycaprolactone; PVC, poly(vinyl chloride); PEG, poly(ethylene glycol); PVDF, poly(vinylidene fluoride); ACP, acrylic copolymer; PEO, poly(ethylene oxide); PET, poly(ethylene terephthalate); PE, polyethylene; PEEK, polyether ether ketone.

ratio of more than 93%. The semicrystalline nature of poly(l-lactide-co-ε-caprolactone)-became completely amorphous while the content of caprolactone increased from 5 to 20 wt%. This can be explained due to the addition of flexible caprolactone destroys regularity of the poly(l-lactide) chains as a consequence inhibits their crystallization. The T_g can be adjusted between 28 and 53 °C by varying the caprolactone content and hard to soft segments ratio. The developed material can be actuated in body temperature and expected to be a prominent candidate for biomedical applications.

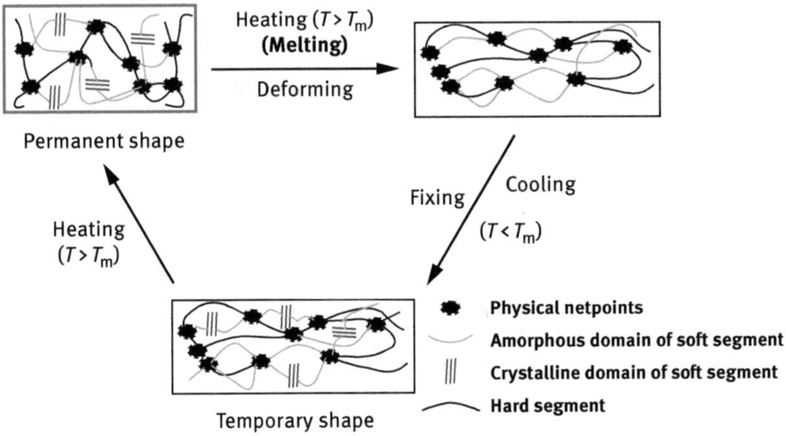

Figure 2.5: The mechanism of SME of physically cross-linked semicrystalline SMP.

2.2.4 Physically cross-linked semicrystalline SMP

In this type of SMP, the soft domains are crystalline and T_m is used as a T_{trans} for SME; thus, the temporary shape is fixed by crystallization of the soft domains. These SMPs generally have hard domains that are responsible for permanent shape and soft domains for temporary shape. The shape memory properties can be modified by changing the hard to soft domain ratio. The mechanism of SME for this type of SMP is shown in Figure 2.5. The polyethylene oxide-co-polyethylene terephthalate, polystyrene-co-poly(butadiene) and the thermoplastic segmented shape memory semicrystalline PUs fall in this category (Table 2.2). Physically cross-linked semicrystalline shape memory PUs synthesized from PEG, MDI and 1,4-butanediol (BD) have been reported [16]. The PEG semicrystalline soft phase acts as a reversible phase, whereas the MDI-BD hard segment serves as physical netpoints. Cyclic, thermomechanical tensile test showed that the material exhibited good shape recovery of more than 80%. Excellent shape fixity can only be achieved when the hard segment content is less than 35 wt%. Kalita and Karak [17] reported the development of PCL-based semicrystalline biobased shape memory PUs. They studied the effect of biobased component (monoglyceride, MG) on the shape memory behaviors of PUs. The shape fixity was found to be increased with the increase of biobased component. They also reported the physically cross-linked semicrystalline biobased hyperbranched shape memory PUs. The SMP was synthesized by reacting PCL as a macroglycol, BD as a chain extender, MG of *Mesua ferrea* L. seed oil as a biobased chain extender, triethanolamine (TEA) as a branch-generating moiety and toluene diisocyanate (TDI) through prepolymerization method using $A_2 + B_3$ approach [18]. The PCL acts as a soft segment and BD-MG-TEA-TDI acts as a hard segment. A series of SMPs have been synthesized using different amounts of branched generating moiety (0–5 wt%). The melting point was increased from 50 to 54 °C with the increase of branched generating moiety content. The hyperbranched PU exhibited higher shape recovery behavior as compared to their linear analog. In addition, SMPs of biodegradable three-arm poly(ε-CL)-based poly(ester-urethanes) actuated at body temperature have been reported [19]. The three-arm PCL triols were synthesized by enzymatic ring-opening polymerization of ε-CL with glycerol. These novel SMPs were synthesized by the reaction of three-arm PCL triols with MDI and 1,6-hexandiol. The SMP exhibited the shape fixity for more than 90% while the shape recovery for more than 95%. T_{trans}, i.e., T_m could be adjusted in the range of 36–39 °C by using PCL triols with T_m of 45–47 °C and *Mn* of 2,720–4,200 g/mol. Shape memory poly(l-lactide-co-ε-caprolactone)-copolymers have been reported. The copolymers exhibited excellent SME with shape recovery and shape fixity ratio of more than 95% by adjusting the compositions. The shape fixity decreased with increase of the ε-CL content. This is due to the decrease of the crystallinity of the copolymer with the increase of the ε-CL content, while the crystalline phase serve as fixing phase. Thermoplastic shape memory PUs were developed by the reaction of poly(tetramethylene glycol) with MDI, and then the chain is extended using different extenders such as linear aliphatic BD, benzoyl-type 4,4-bis (4-hydroxyhex-oxy)-isopropylane and naphthalate-type bis(2-phenoxyethanol)-sulfone or naphthoxy diethanol [20]. T_g of these copolymers was in the range of −73 to (−)50 °C for the soft segment (T_{gs}) and 70–106 °C for the hard segment (T_{gh}), and T_m was in the range of 14–24 °C for the soft segment and 198–206 °C for the hard segment. The materials exhibited shape recovery and shape fixity ratio above 90%.

2.3 Based on external stimulus

2.3.1 Thermoresponsive SMP

In this class of SMP, the polymer is actuated by direct heating [10, 20–22]. By far these are the most common SMPs. Thermoresponsive SMPs recover their original shape when these are heated above T_g for amorphous polymers whereas heated above T_m for crystalline polymers. When the polymer is heated above T_g, the polymer chains start to vibrate and slip of each other at which the storage energy of the system is released, and as a result SMP will return to their original position. An entropy loss arises when the polymer is deformed, which corresponds to entropy elasticity, whereas in the rubbery state the polymer tends to recover its original shape through recovering the entropy loss. Kalita and Karak [23] reported the thermoresponsive biobased shape memory PUs. They prepared series of shape memory PUs using different vegetable oils such as castor oil, sunflower oil and *Mesua ferrea* L. oil. PUs showed T_m in the range of 39–43 °C. All the shape memory PUs exhibited good shape recovery and shape fixity behavior under heat actuation at a temperature of 60 °C. The castor oil-based one showed the faster shape recovery among the others. This is attributed to the increase in stored energy as a result of more physical cross-links and the presence of more functionality in the castor oil. They also reported the thermoresponsive PU/multiwalled carbon nanotube (MWCNT) shape memory nanocomposites [24]. The shape recovery was found to be increased with the increase of MWCNT content due to the increase in strain energy and modulus of the system. Again the author studied the shape memory behaviors of nanocomposites in hot water [25]. The material showed faster shape recovery in hot water as compared to the normal heating. This can be attributed to the higher heat conducting nature of water than air. Lee et al. [26] studied the thermoresponsive shape memory behaviors of PU block copolymers. They studied the SME on different weight percentages of hard segment content in PU. All PUs showed more than 90% shape fixity at each hard segment content. However, shape recovery was not noticed at 20, 25 and 50 wt% of hard segment content. PUs containing 20 or 25 wt% of hard segment cannot have strong interaction or physical cross-links such as dipole–dipole interaction, hydrogen bonding and induced dipole–dipole interaction because of the low content of hard segment, resulting in the loss of SME. PU with 50 wt% of hard segment did not exhibit shape recovery because of the excess interactions among the hard segments leading to a rigid structure. PUs with 30–45 wt% of hard segment content showed 80–95% of shape recovery, where PU copolymers can make strong interactions among hard segments that are enough to restore the polymer to revert to their original shape. Reyntjens et al. [27] reported the thermoresponsive poly(octadecyl vinyl ether)-co-butyl acrylate SMP. The material exhibited shape recovery within seconds when heating at 60 °C. The extensive phase separation occurring in these materials is attributed to the block structure of the networks, where the octadecyl vinyl ether units are allowed to crystallize, resulting in the good SME.

2.3.2 Light-responsive SMP

Light-responsive SMP can be achieved either via photochemistry or through light-induced heating. In this SMP, light is induced to the SME, which has some photoresponsive groups that act as molecular switches [22, 28–30]. This stimulation does not have any relation with the temperature effect. However, in the case of light-induced heating SME, the polymer and/or nanofillers absorb the irradiated light and generate the induced heating resulting in the SME. SMPs are stretched and irradiated by certain light of wavelength and consequently the photoresponsive groups form cross-links. SMPs obtain a new shape and retain the shape even when releasing the stress. When this temporary shape is irradiated by certain light of wavelength, the cross-links cleave allowing the SMP to recover its original shape. Lendlein et al. [31] reported the UV light-responsive SMP. They fixed the elongated polymer applying the UV light at wavelength higher than 260 nm and recovered the original shape using wavelength of light shorter than 260 nm at room temperature. Light-induced pendant cinnamamide groups containing polyester-PU SMP have been developed [32]. The material exhibited excellent shape fixity and shape recovery under the UV stimulus at ambient temperature. The pendant cinnamamide groups serve as photoresponsive molecular switches and provide the polymer with reversible [2 + 2] cycloaddition cross-linking. The temporary shape was obtained by irradiating with >260 nm UV light after deforming via an external force, as a result of the formation of temporary chemical cross-links through a [2 + 2] cycloaddition reaction. The original shape was recovered after irradiation with <260 nm UV light through cleaving of the light-induced cross-links (Figure 2.6). Infrared (IR) light stimuli carbon nanoparticles filled with

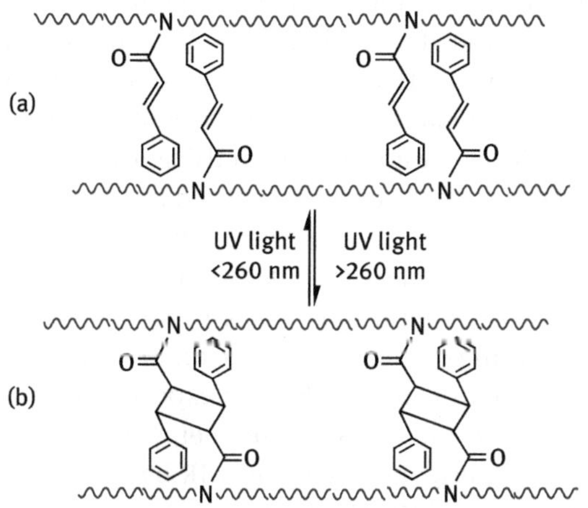

(a) PCL segments containing cinnamamide moieties

(b) Light-induced reversibly cross-linked soft segments

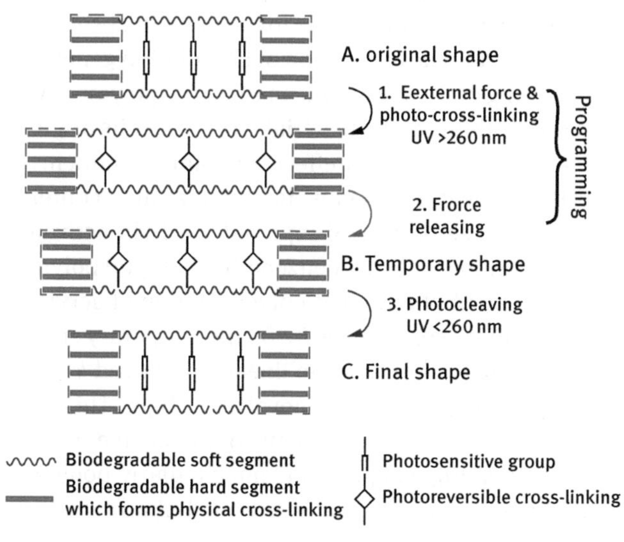

Figure 2.6: SME of UV light-induced SMP.

styrene-based SMP have been reported [33]. The material exhibited strong SME under the influence of IR light. The IR light absorption capacity of carbon nanoparticles induces internal heat and that triggers the SMP to recover its original shape. Medium IR laser light-activated optical fiber-embedded styrene-based SMPs have also been developed. The material exhibited faster excellent shape recovery behaviors. When the IR light activates, the absorbed optical energy of the polymer is transformed into thermal energy by Joule heating. Once the temperature approaches to T_{trans}, the deformed SMPs recover to its original shape. Photoresponsive metal–ligand supramolecular SMPs have been developed [34]. Metallosupramolecular polymers were prepared by the end-capped 4-oxy-2,6-bis(*N*-methylbenzimidazolyl) pyridine with low-molecular-weight poly(butadiene) followed by the addition of metal salts and cross-linked with tetrathiol. The material exhibited more than 90% shape fixity and shape recovery behaviors. Fixed shape can be obtained by the photo-cross-linking of the poly(butadiene) core with a tetrafunctional thiol through a photo-induced thiol-ene reaction. The metal–ligand complexes are able to absorb UV light, and this absorbed energy is converted into fluorescence and leads to a localized heating. This localized heating renders the softening of the hard phase and increased decomplexation or rate of exchange of the metal–ligand complexes. Thus, the material can be deformed while irradiating with UV light (320–390 nm), and upon removal of the light the sample becomes cool, the metal–ligand interactions re-engage and reforms the phase separation, locking in the temporary deformed state. Further exposure to light will again break up the phase separation, and the stored strain energy is released resulting in return to its entropically favored, original shape. Near-IR light-induced PU/graphene oxide hybrids have been developed [35]. The modified graphene oxide acts as a multifunctional cross-linker at low contents and a nucleating agent at high contents and IR absorbing materials. The hybrid material exhibited more than 90% shape recovery behaviors under the actuation of IR radiation. The recovery speed increased with the increase of graphene oxide content due to the more IR absorption and increase in conductivity of the material. The absorption of IR radiation generates induced heating that leads to the melting of the crystalline mass of the polymer. As a consequence, there is decrease in the modulus of the material and recover its original shape. At low content of graphene oxide, cross-linker effect is dominant inducing a high degree of recovery, whereas the nucleating effect is dominant at high content, giving high ΔH_m and high shape fixity. Lee et al. [36] reported the light-activated azobenzene-based liquid polymer crystalline SMP. Azobenzene-based liquid polymer crystalline networks were synthesized by photoinitiated polymerization of acrylate-based monomer and azobenzene containing cross-linker. To fix the temporary shape, the material is deformed to bending shape and exposed to linearly polarized 442 nm laser light. The material retains its bending shape after removal of the light. This photofixed shape is because of the light-directed rearrangements to the polymer chains in the glassy matrix. By exposing the bending shape to 442 nm circularly polarized light, it recovers to the original shape. A new optically triggered shape memory composite material was developed from poly(3-caprolactone) cross-linked with hexamethylene diisocyanate and surface-functionalized gold (Au) nanoparticles [37]. The composites exhibited the good shape recovery behavior under the laser actuation. Gold nanoparticles between 60 and 200 nm in size have a strong absorption of light especially IR light. The irradiated light can be absorbed by the gold nanoparticles to enhance the temperature of gold nanoparticles/SMP composites locally to actuate the shape recovery. However, the shape recovery process could be stopped at any stage by turning off the laser source, and then resumed at anytime by irradiating the laser on sample. Kalita and Karak [38] reported the microwave-actuated PU-Fe_3O_4 SMP nanocomposites. They reported that the composites showed almost full shape recovery under the microwave stimulus. No significant change of shape recovery of the nanocomposites was obtained over the repeated cycles (Ten) of test. When the microwave is irradiated on the sample, the dipole moment of the material oscillates to align with the oscillating external electric field; as a consequence, heat is generated because of the molecular friction and collisions. The shape recovery was observed when the induced heat is close to the T_{trans} and found to increase with the increase of Fe_3O_4 content under the microwave irradiation. The shape recovery time decreased (81–49 s) with the increase of Fe_3O_4 nanoparticles content in the nanocomposites. The shape recovery speed may be increased because of the enhanced microwave absorption characteristic of the nanocomposites with the increase of Fe_3O_4 nanoparticles content.

2.3.3 pH-responsive SMP

In this category of SMP, materials respond to the change of pH. In the presence of ionizable functional groups, the polymer gets ionized and acquired charge in certain pH [39]. The polymer chains between like-charged groups lead to repulsion and hence expand their dimensions. When the pH changed, the repulsion is gone and the materials come to the original position. A novel pH-responsive SMP nanocomposite has been developed by blending PEG–poly(ε-CL)-based PU with functionalized cellulose nanocrystals (CNCs) [40]. CNCs were functionalized with carboxyl groups (CNC–CO_2H) through 2,2,6,6-tetramethyl-1-piperidinyloxy and with pyridine moieties (CNC–$C_6H_4NO_2$) through hydroxyl substitution of CNCs with pyridine-4-carbonyl chloride, respectively. At a high pH value, the CNC–$C_6H_4NO_2$ had attractive interactions from the hydrogen bonding between pyridine groups and hydroxyl moieties. However, at a low pH value, the interactions reduced or disappeared because of the protonation of pyridine moieties. The excellent pH-actuated SME can be readily controlled through changing hydrogen bonding interaction by varying the pH of the environment. A pH-sensitive PEG-based PU triple SMP has been reported [41]. The association and disassociation of carboxylic dimers present in the polymer chains act as two switches to control the triple SME. The carboxylic dimer is affected by pH values to associate in acidic solutions (at pH 2), whereas dissociate in alkaline solutions (at pH 9) to induce the pH-responsive SME. The carboxylic dimers play a vital role in the formulation of pH-responsive shape memory PU. A novel pH-responsive SMP is developed by cross-linking the β-cyclodextrin-modified alginate (β-CD-Alg) and diethylenetriamine-modified alginate (DETA-Alg) [42]. Temporary shape of the material can be fixed at pH 11.5 and recover its original shape at pH 7. At pH 11.5, the material was rigid and could not deform freely due to the presence of both fixing cross-links (alginate chains and calcium cations) and reversible cross-links (β-CD-DETA). At pH 7, the protonation of amino of DETA causes the dissociation of β-CD-DETA. As a consequence, the material would have become soft and readily be deformed under the application of an external stress. At pH 11.5 under an external stress, the deformation of the specimen would be frozen and render a temporary shape. Once the specimen was transferred at pH 7, the dissociation of β-CD-DETA resulted in disappearance of the strength of the frozen deformed specimen. As a result, the material recovered to its original shape. Chen et al. [43] reported a highly pH-responsive SMP, which was synthesized by introducing pyridine rings into the PU backbone. The mechanism of pH responsiveness is the formation of a hydrogen bond interaction between H–N of urethane and the N atom of the pyridine ring in neutral or alkaline condition which is dissociated under acidic conditions due to the protonation of the pyridine ring.

2.3.4 Solvent-responsive SMP

Water or solvent can be used as a stimulus for SME of SMP. The solvent molecules penetrate to the bulk of the polymer and depress the T_{trans} by means of softening, swelling or dissolving the switching domain [44–46]. T_{trans} may drop below the room temperature and this leads to the SME of solvent-responsive SMP. SME depends on the solvent molecule size, shape, polarity, solubility parameter and so on. Solvent-induced chemically cross-linked PVA SMPs have been reported [3]. Shape memory behaviors were studied in good or poor solvents (including water, dimethylformamide [DMF] and ethylene glycol) for PVA in room temperature. The chemical cross-linked PVA exhibited excellent shape fixity ratio and recovery ratio in water even after five cycle applications. The solubility parameter and polarity of solvents are the most important factors influencing the sorption and diffusion of small molecules in the polymer, which determines the degree of swelling and macroscopic shape recovery. PVA swelling leads to the decrease of T_g and increase in the flexibility of chain segments. The high chain mobility allows the frozen stress to release, resulting in macroscopic shape recovery. Solvent-induced styrene-based thermosetting SMPs have been developed [47]. The material exhibited excellent shape recovery in DMF. The absorption of solvent in the SMP weakens the elasticity modulus, which causes a significant

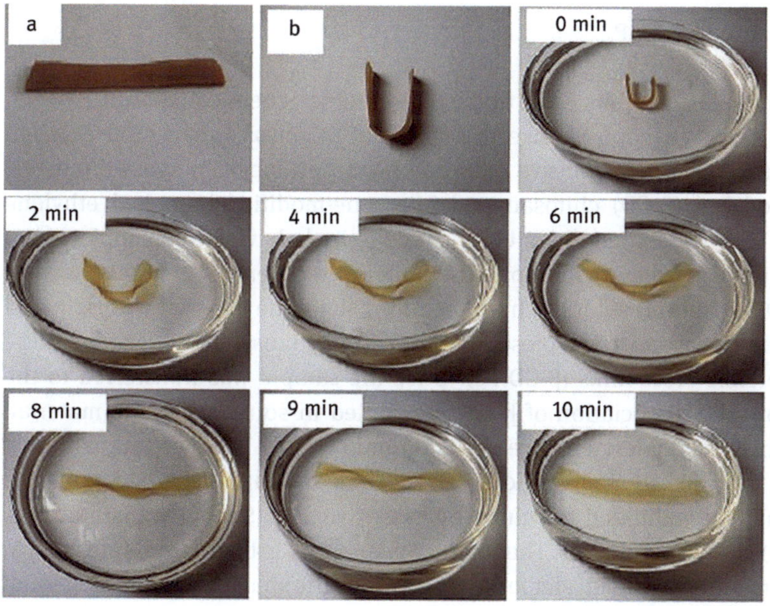

Figure 2.7: SME of solvent-induced SMP: (a) original shape and (b) temporary shape.

decrease in T_g, which has a direct influence on the shape recovery of SMP. Kalita et al. [48] studied the solvent-induced shape memory behaviors of biobased PUs. They studied the shape memory behaviors in different solvents. In DMF the material exhibited excellent shape recovery in shorter recovery time within 10 s (Figure 2.7). The solvent molecules penetrate into the polymer chains after immersing the fixed sample (temporary shape) in the solvent, and decrease the transition/switching temperature of the polymer. When the transition/switching temperature of the switching segments is lower than the temperature of their surroundings, active motion of the switching segments occurs. As a consequence, release the stored strain energy in the temporary fixed shape sample, and hence the sample transforms back to its original shape. The solubility parameter, polarity and size of the solvent molecules are the factors that exert the strongest influence on the shape recovery behavior of the polymer. Water-responsive cellulose nanowhisker (CNW)/TPU nanocomposite SMPs have been developed [49]. SMPs showed excellent shape recovery in shorter period by wetting in water immersion (10 minutes). Wetting of the sample (original shape) can soften the CNW/TPU, as water molecules break up the hydrogen bonds between the nanowhiskers. This allows the material for easy deformation into a temporary shape. The subsequent drying results in shape fixation through the formation of a hydrogen-bonded three-dimensional network of individualized whiskers through removal of the water molecules. In the recovery stage, wetting as the external stimulus leads to dissociation of the network of whiskers and actuates the shape recovery of its original shape.

2.3.5 Chemoresponsive SMP

Chemoresponsive SMP exhibits SME under the external chemical stimuli [50–52]. Hydrolysis and redox reaction change the chemical or physical states on external chemical stimuli. A chemoresponsive SMP has been developed by cross-linking β-cyclodextrin-modified (β-CD) chitosan and ferrocene-modified branched ethylene imine polymer [53]. The material exhibited shape fixity behavior when subjected to a 1% $NaHSO_3$ (reducing agent) solution. The material recovers its original shape in 1% $(NH_4)Ce(NO_3)_6$ (oxidant) solution. In the reduced state, the material was rigid and could not deform freely due to the presence of both fixing cross-links (covalent cross-links) and reversible cross-links (β-CD-Fc). Once the sample was transferred to the oxidant solution, the dissociation of β-CD-Fc resulted in softening of the material. As a consequence, the material recovered its original shape. The material exhibited shape recovery ratio and the fixity ratio above 70%. Incorporation of glucose oxidase (GOD) in the material exhibited the glucose-responsive SMP. Similarly, material was fixed into a 0.2% $NaHSO_3$ solution. The temporary shape was transferred into a 0.2 M glucose solution and the original shape was recovered. The shape memory behavior of the material was caused by hydrolysis of glucose catalyzed by GOD. Sugar-responsive phenylboronic acid-grafted alginate (Alg-PBA)/PVA shape memory hydrogel have been developed [54]. Alg-PBA and PVA were mixed under basic conditions, and supramolecular hydrogels could be formed due to the formation of dynamic covalent PBA-diol ester bonds between PBA groups of Alg-PBA and hydroxyl groups of PVA. Then, the prepared Alg-PBA/PVA hydrogels were immersed into $CaCl_2$ aqueous solution and the alginate chains were cross-linked by Ca^{2+}. There are two types of cross-links inside the hydrogels: the Alg-Ca^{2+} cross-link was responsible to determine the permanent shape, whereas dynamic PBA-diol ester bonds serve as reversible cross-link and memorize the temporary shape. When the temporary shape of the hydrogel is immersed into a glucose solution, it recovers the original shape. The interaction between PBA groups and monosaccharides such as glucose and fructose is stronger than that between PBA groups and PVA. Thus when a fixed temporary shape was immersed in aqueous solution of glucose and fructose, the reversible PBA-diol ester bonds between Alg-PBA and PVA will be dissociated, which drives the shape recovery of the hydrogel. The material exhibited excellent shape fixity and shape recovery to almost 100%.

2.3.6 Electroresponsive SMP

In this class of SMP, SME is observed under the application of electrical current/voltage [22, 46, 55]. Most of the SMPs are insulating in nature. These properties prevent them from the actuation of electrical resistance effect. This type of SMP can be prepared by incorporation of carbon nanotube, graphene, carbon nanofibers, carbon black, metal nanoparticles, polypyrrole and so on as conducting fillers. The electric current passes through the conductive filler network in the SMP; the internal temperature of the SMP increases through Joule heating phenomena and once its temperature is above T_{trans} it activates the SMP [56, 57]. Cho et al. [58] first reported the SME of SMP nanocomposites using CNT as conducting filler by electric field. MWCNTs were used after chemically surface modification in a mixed solvent of nitric acid and sulfuric acid, to increase interfacial bonding between polymers and MWCNTs to obtain conducting SMP. The electrical conductivity was found in the order of 10^{-3} S cm for the 5 wt% MWCNT-based nanocomposite. The material showed excellent SME, and the surface temperature of the nanocomposites was found to be above 35 °C in 8 s when the applied voltage was 40 V. Du et al. [59] developed electroactive SMP based on PVA/MWCNT composites. The volume electrical resistivity of pure PVA is 2.2×10^{14} Ω cm. The electrical percolation network was formed above 1 wt% because of low aspect ratio of used MWCNT. When the loading amount reached 10 wt%, the electrical resistivity drastically lowered to 1.4×10^{3} Ω cm, indicating that the MWCNT connect to each other and form a better electrical conductive network in the PVA matrix. However, the rate of decrease of electrical resistivity of the nanocomposites became constant when the MWCNT content increased from 10 to 30 wt%. The SMP PVA/MWCNT (20 wt%) nanocomposite recovered its original shape within 35 s, when a constant voltage of 60 V was applied. However, for the 10 and 15 wt% samples, higher voltages of 120 and 110 V were required to recover its original shape, respectively. The PVA/MWCNT composites are in a semiconductor or insulator state that could not be driven by voltage in the range from 0 to 120 V when the concentration of MWCNT is <10 wt%. However, in the case of PVA/MWCNT composites with 30 wt% of MWCNT, the initial recovery time increased because the transport of the Joule heating slowed down as the thermal conductivity decreased due to the aggregation of nanotubes. Low-temperature electroactive shape memory PU–SWCNT hybrids have been developed [60]. The conductivity of PU–SWCNT hybrids was 4.86×10^{-7} S/cm with an addition of 1% SWCNTs, but their conductivity enhanced to 5.9×10^{-3} S/cm with an addition of 2 wt% SWCNTs. The conductivity of PU–SWCNT hybrids continued to increase to 0.2677 S/cm with an addition of 5 wt% SWCNT. The result suggested that critical filler concentration was required to reach the true electrical conductivity known as the percolation threshold. The composite with 4 wt% SWCNT possesses the optimum electroactive SME with shape recovery to about 88% within 90 s under the applied voltage of 30 V. An electroactive hydroepoxy/carbon black composite has been reported [56]. The percolation threshold of the electroactive hydroepoxy/carbon black shape memory composite is lower than that of many other composites. The electroactive hydroepoxy composite exhibited excellent SME because of the low percolation threshold. The composite filled with 1.9 wt% carbon black recovered its original shape to nearly 100% in only a few minutes under an applied voltage of 200 V. The study revealed that the shape recovery time decreased with the increase of applied voltage. According to Joule's law, the Joule heating is related to the voltage such that increased Joule heating would be induced with an increase in the voltage. As a result, as the voltage increases, the temperature of the sample simultaneously increases, rendering a higher recovery rate. Electroactive shape memory composites were developed using PU block copolymer

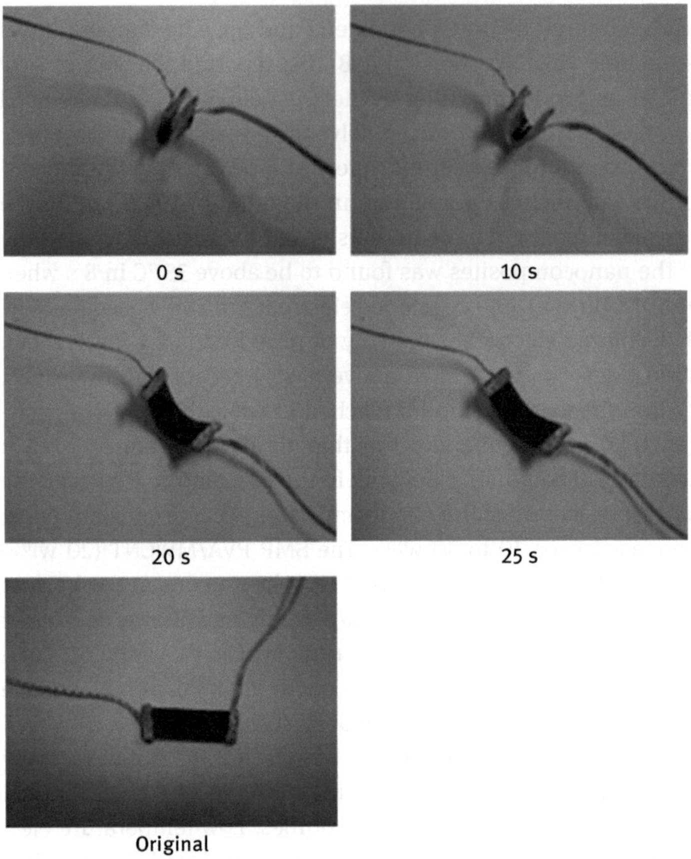

Figure 2.8: SME of electroresponsive SMP at the voltage of 40 V.

and conducting polypyrrole by chemical oxidative polymerization [61]. The conductivity of the composites increased with increase of polypyrrole, and a high conductivity of the order of 10^{-2} S/cm was obtained at 6–20 wt% polypyrrole. Such a conductivity of the composites was sufficient to show electroactive SME. The material exhibited shape recovery of 85–90% in 20–25 s when an electric field of 40 V was applied (Figure 2.8).

2.3.7 Magnetoresponsive SMP

This type of SMP can be actuated by the alternating magnetic field [45, 62, 63]. SME can be achieved by inclusion of fillers such as ferromagnetic particles, iron oxide nanoparticles and nickel powders into the SMP. They offer the opportunity to actuate the SMP through contactless by a specific remote control. The ferromagnetic particles act as inductive heater under the application of alternating magnetic field through hysteresis loss, eddy current loss and other mechanisms based on the different kinds of magnetic particles and their sizes [64, 65]. The amount of heat generated by magnetic particles depends on the size of the particles and frequency of the applied magnetic field. The shape recovery of the composites can be actuated by means of inductive Joule heating in an alternating magnetic field as a consequence of transforming electromagnetic energy from an external high-frequency field to heat due to relaxation processes. Sometimes it is difficult to control the amount of heat generated by the magnetic particles in SMP, which is often downside for biomedical applications. Some ferromagnetic particles incorporated in the SMP enable innate thermoregulation, which was caused by the Curie temperature (T_c). T_c is the temperature at which a ferromagnetic material becomes paramagnetic, losing its ability to generate heat through a hysteresis loss mechanism. Magnetic particles with appropriate sizes will generate heat primarily through a magnetic hysteresis loss mechanism instead of an eddy current mechanism. It becomes possible to have an innate thermoregulation mechanism that limits the maximum attainable temperature to T_c. By choosing magnetic materials with suitable T_c could significantly eliminate the danger of overheating in clinical applications.

Cross-linked poly(ε-CL)/Fe_3O_4 magnetically induced SMP have been reported [66]. The materials showed excellent shape recovery in an alternating magnetic field with $f = 20$ kHz and $H = 6.8$ kA/m. SME is observed in an alternating magnetic field because Brownian and Néel relaxation losses of Fe_3O_4 can be transferred into heat. The shape recovery time was 130 s when the specimen was inductively heated in the alternating magnetic field. It was suggested that faster recovery speed can be achieved by increasing the frequency and employing a heat shield for specimens to prevent the heat loss during inductively heated. Mohr et al. [67] studied that magnetically actuated Fe_2O_3 nanoparticles incorporated thermoplastic polyether urethane nanocomposites. Iron oxide nanoparticles with average diameters at 20–30 nm exhibited homogeneous distribution within the SMP matrix. The shape recovery of SMP composites was investigated by inductive heating in an alternating magnetic field ($f = 258$ kHz; $H = 30$ kA/m). The sample geometry and nanoparticle content have an important role to achieve maximum temperatures by inductive heating in a specific magnetic field. Schmidt [62] incorporated surface-modified Fe_3O_4 (diameter: 11 nm) superparamagnetic nanoparticles into an SMP matrix. She reported that magnetically actuated thermosetting SMP composite of oligo(ε-CL)dimethacrylate/butyl acrylate contains Fe_3O_4 magnetite nanoparticles (2–12 wt%). The specific power loss of the particles was found to be 30.8 W/g at 300 kHz with a power of 5.0 kW. The material recovers its original shape in about 20 s with the application of an AC field of 300 kHz (Figure 2.9). A maximum sample temperature of 42 °C was found during inductive heating. Nickel zinc ferrite particles have been reported to achieve SMP actuation through inductive heating by means of innate thermoregulation. T_c of this material could be varied in a wide range by changing the composition. Buckley et al. [64] reported that nickel zinc ferrite particles incorporated magnetically actuated SMP for biomedical applications. The effect of volumetric particle loading from 1% to 20% on the heating efficiency of the magnetic particles was also investigated. The heating efficiency of the particles found to be decreased with

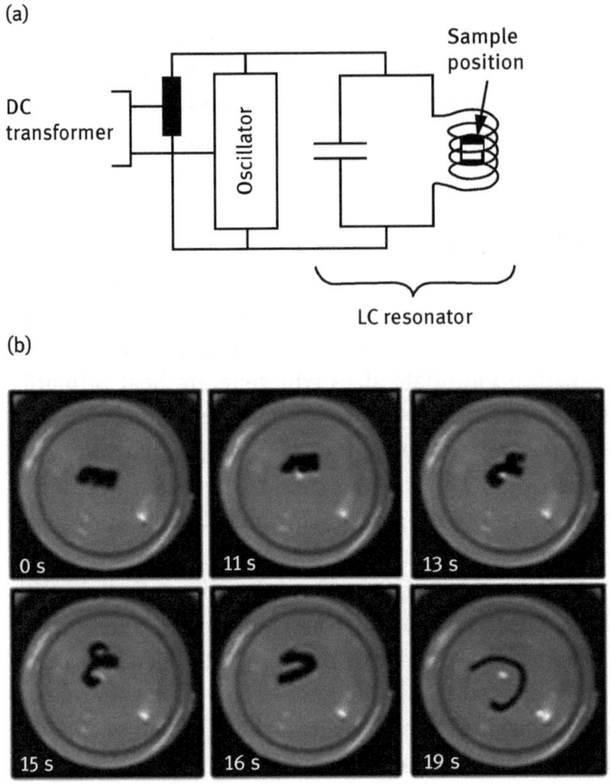

Figure 2.9: SME of magnetoresponsive SMP: (a) schematic diagram of the LC resonant circuit-based HF generator used for induction heating experiments, and sample positioning and (b) shape recovery under the impact of HF electromagnetic field.

the increase of the volume content of the particles due to the magnetic shielding effect as a consequence of agglomeration of particles in SMP matrix. This magnetic shielding effect could lower the volumetric power dissipation due to lower magnetic field strength experienced by the particles. The addition of 10% volume content of particles provides sufficient heating for SMP actuation in air using magnetic field ($f = 12.2$ MHz and $H = 545$ A/m) and does not interfere significantly with the SME of SMP.

References

[1] Liu C, Qin H, Mather PT. Review of progress in shape-memory polymers. J Mater Chem 2007;17(16):1543–58.

[2] Hager MD, Bode S, Weber C, Schubert US. Shape memory polymers: past, present and future developments. Prog Polym Sci 2015;49–50:3–33.

[3] Du H, Zhang J. Solvent induced shape recovery of shape memory polymer based on chemically cross-linked poly(vinyl alcohol). Soft Matter 2010;6(14):3370–76.

[4] Zhou J, Schmidt AM, Ritter H. Bicomponent transparent polyester networks with shape memory effect. Macromolecules 2010;43(2):939–42.

[5] Chun BC, Cha SH, Park C, Chung Y, Park MJ, Whan J. Dynamic mechanical properties of sandwich-structured epoxy beam composites containing poly (ethyleneterephthalate)/ poly(ethylene glycol) copolymer with shape memory effect. J Appl Polym Sci 2003;90(11):3141–9.

[6] Rousseau IA, Xie T. Shape memory epoxy: composition, structure, properties and shape memory performances. J Mater Chem 2010;20(17):3431–41.

[7] Alteheld A, Feng Y, Kelch S, Lendlein A. Biodegradable, amorphous copolyester-urethane networks having shape-memory properties. Ang Chem Int Ed 2005;44(8):1188–92.

[8] Yakacki CM, Shandas R, Safranski D, Ortega AM, Sassaman K, Gall K. Strong, tailored, biocompatible shape-memory polymer networks. Adv Funct Mater 2008;18(16):2428–35.

[9] Liu C, Chun SB, Mather PT, Zheng L, Haley EH, Coughlin EB. Chemically cross-linked polycyclooctene: synthesis, characterization, and shape memory behavior. Macromolecules 2002;35(27):9868–74.

[10] Hu J, Yang Z, Yeung L, Ji F, Liu Y. Crosslinked polyurethanes with shape memory properties. Polym Int 2005;54(5):854–9.

[11] Kolesov IS, Kratz K, Lendlein A, Radusch HJ. Kinetics and dynamics of thermally-induced shape-memory behavior of crosslinked short-chain branched polyethylenes. Polymer 2009;50(23):5490–8.

[12] Paderni K, Pandini S, Passera S, Pilati F, Toselli M, Messori M. Shape-memory polymer networks from sol-gel cross-linked alkoxysilane-terminated poly(ε-caprolactone). J Mater Sci 2012;47(10):4354–62.

[13] Kelch S, Steuer S, Schmidt AM, Lendlein A. Shape-memory polymer networks from oligo[(ε-hydroxycaproate)-co-glycolate]dimethacrylates and butyl acrylate with adjustable hydrolytic degradation rate. Biomacromolecules 2007;8(3):1018–27.

[14] Cho JW, Jung YC, Chung YC, Chun BC. Improved mechanical properties of shape-memory polyurethane block copolymers through the control of the soft-segment arrangement. J Appl Polym Sci 2004;93(5):2410–15.

[15] Jeong HM, Ahn BK, Kim BK. Miscibility and shape memory effect of thermoplastic polyurethane blends with phenoxy resin. Eur Polym J 2001;37(11):2245–52.

[16] Mo F, Zhou F, Chen S, Yang H, Ge Z, Chen S. Development of shape memory polyurethane based on polyethylene glycol and liquefied 4,4'-diphenylmethane diisocyanate using a bulk method for biomedical applications. Polym Int 2015;64(4):477–85.

[17] Kalita H, Karak N. Bio-based elastomeric hyperbranched polyurethanes for shape memory application. Iran Polym J 2012;21(4):263–71.

[18] Kalita H, Karak N. Mesua ferrea L. seed oil-based hyperbranched shape memory polyurethanes: effect of multifunctional component. Polym Eng Sci 2012;52:2454–61.

[19] Xue L, Dai S LZ. Synthesis and characterization of three-arm poly(3-caprolactone)-based poly (ester urethanes) with shape-memory effect at body temperature. Biomaterials 2009;42:964–72.

[20] Wang HH, Yuen UE. Synthesis of thermoplastic polyurethane and its physical and shape memory properties. J Appl Polym Sci 2006;102(1):607–15.

[21] Mondal S. Recent developments in temperature responsive shape memory polymers. Mini Rev Org Chem 2009;6(2):114–19.

[22] Meng H, Jinlian Hu. A brief review of stimulus-active polymers responsive to thermal, light, magnetic, electric, and water/solvent stimuli. J Intell Mater Syst Struct 2010;21(9):859–85.

[23] Kalita H, Karak N. Biobased hyperbranched shape-memory polyurethanes: effect of different vegetable oils. J Appl Polym Sci 2014;131(1):1–8.

[24] Kalita H, Karak N. Bio-based hyperbranched polyurethane/multi-walled carbon nanotube nanocomposites as shape memory materials. Polym Compos 2014;35(4):636–43.

[25] Kalita H, Karak N. Bio-based hyperbranched polyurethane / multi-walled carbon nanotube nanocomposites as shape memory materials. J Nanoeng Nanomanufacturing 2013;3(7):194–201.

[26] Lee BS, Chun BC, Chung Y-C, Sul K II, Cho JW. Structure and thermomechanical properties of polyurethane block copolymers with shape memory effect. Macromolecules 2001;34(18):6431–8.

[27] Reyntjens WG, Du Prez FE, Goethals EJ. Polymer networks containing crystallizable poly(octadecyl vinyl ether) segments for shape-memory materials. Macromol Rapid Commun 1999;20(5):251–5.

[28] Behl M, Lendlein A. Shape-memory polymers. Mater Today 2007;10(4):20–8.

[29] Iqbal D, Samiullah MH. Photo-responsive shape-memory and shape-changing liquid-crystal polymer networks. Materials (Basel) 2013;6(1):116–42.

[30] Beblo R V., Weiland LM. Demonstration of a multiscale modeling technique: prediction of the stress–strain response of light activated shape memory polymers. Smart Mater Struct 2010;19:94012–20.

[31] Lendlein A, Jiang H, Jünger O, Langer R. Light-induced shape-memory polymers. Nature 2005;434:879–82.

[32] Wu L, Jin C, Sun X. Synthesis, properties, and light-induced shape memory effect of multiblock polyester urethanes containing biodegradable segments and pendant cinnamamide groups. Biomacromolecules 2011;12(1):235–41.

[33] Leng J, Wu X, Liu Y. Infrared light-active shape memory polymer filled with nanocarbon particles. J Appl Polym Sci 2009;114:2455–60.

[34] Kumpfer JR, Rowan SJ. Thermo-, photo-, and chemo-responsive shape-memory properties from photo-cross-linked metallo-supramolecular polymers. J Am Chem Soc 2011;133(32):12866–74.

[35] Park JH, Kim BK. Infrared light actuated shape memory effects in crystalline polyurethane/graphene chemical hybrids. Smart Mater Struct 2014;23:1–7.

[36] Lee KM, Koerner H, Vaia RA, Bunning TJ, White TJ. Light-activated shape memory of glassy, azobenzene liquid crystalline polymer networks. Soft Matter 2011;7(9):4318–24.

[37] Zhang H, Xia H, Zhao Y. Optically triggered and spatially controllable shape-memory polymer–gold nanoparticle composite materials. J Mater Chem 2012;22(3):845–9.

[38] Kalita H, Karak N. Bio-based hyperbranched polyurethane/Fe_3O_4 nanocomposites as shape memory materials. Polym Adv Technol 2013;24(9):819–23.

[39] Ahn S, Kasi RM, Kim S-C, Sharma N, Zhou Y. Stimuli-responsive polymer gels. Soft Matter 2008;4(6):1151–57.

[40] Li Y, Chen H, Liu D, Wang W, Liu Y, Zhou S. PH-responsive shape memory poly(ethylene glycol)-poly(ε-caprolactone)-based polyurethane/cellulose nanocrystals nanocomposite. ACS Appl Mater Interfaces 2015;7(23):12988–99.

[41] Song Q, Chen H, Zhou S, Zhao K, Wang B, Hu P. Thermo- and pH-sensitive shape memory polyurethane containing carboxyl Groups. Polym Chem 2016;7:1739–46.

[42] Han XJ, Dong ZQ, Fan MM. pH-induced shape-memory polymers. Macromol Rapid Commun 2012;33(12):1055–60.

[43] Chen H, Li Y, Liu Y, Gong T, Wang L, Zhou S. Highly pH-sensitive polyurethane exhibiting shape memory and drug release. Polym Chem 2014;5(17):5168–74.

[44] Lv H, Leng J, Liu Y, Du S. Shape-memory polymer in response to solution. Adv Eng Mater 2008;10(6):592–5.

[45] Leng J, Lan X, Liu Y, Du S. Shape-memory polymers and their composites: stimulus methods and applications. Prog Mater Sci 2011;56(7):1077–135.

[46] Meng H, Li G. A review of stimuli-responsive shape memory polymer composites. Polymer 2013;54(9):2199–221.

[47] Lu H, Liu Y, Leng J, Du S. Qualitative separation of the physical swelling effect on the recovery behavior of shape memory polymer. Eur Polym J 2010;46(9):1908–14.

[48] Kalita H, Mandal M, Karak N. Biodegradable solvent-induced shape-memory hyperbranched polyurethane. J Polym Res 2012;19(10):9982–90.

[49] Zhu Y, Hu J, Luo H, Young RJ, Deng L, Zhang S, et al. Rapidly switchable water-sensitive shape-memory cellulose/elastomer nano-composites. Soft Matter 2012;8(8):2509–17.

[50] Huang WM, Zhao Y, Zhang JL, Lu HB. Chemo-responsive shape memory/change effect in polymeric materials based on transport phenomena. J Fluid Flow Heat Mass Transfer 2014;1:16–22.

[51] Lu HB, Huang WM, Yao YT. Review of chemo-responsive shape change/memory polymers. Pigment Resin Technol 2013;42(4):237–46.

[52] Lu H, Leng J, Du S. A phenomenological approach for the chemo-responsive shape memory effect in amorphous polymers. Soft Matter 2013;9(14):3851–58.

[53] Dong ZQ, Cao Y, Yuan QJ, Wang YF, Li JH, Li BJ, et al. Redox- and glucose-induced shape-memory polymers. Macromol Rapid Commun 2013;34(10):867–72.

[54] Meng H, Zheng J, Wen XF, Cai ZQ, Zhang JW, Chen T. PH- and sugar-induced shape memory hydrogel based on reversible phenylboronic acid-diol ester bonds. Macromol Rapid Commun 2015;36(6):533–7.

[55] Liu Y, Lv H, Lan X, Leng J, Du S. Review of electro-active shape-memory polymer composite. Compos Sci Technol 2009;69(13):2064–68.

[56] Wei K, Zhu G, Tang Y, Li X. Electroactive shape-memory effects of hydro-epoxy/carbon black composites. Polym J 2013;45(8):871–6.

[57] Luo X, Mather PT. Conductive shape memory nanocomposites for high speed electrical actuation. Soft Matter 2010;6(10):2146–49.

[58] Cho JW, Kim JW, Jung YC, Goo NS. Electroactive shape-memory polyurethane composites incorporating carbon nanotubes. Macromol Rapid Commun 2005;26(5):412–16.

[59] Du FP, Ye EZ, Yang W, Shen TH, Tang CY, Xie XL, et al. Electroactive shape memory polymer based on optimized multi-walled carbon nanotubes/polyvinyl alcohol nanocomposites. Compos Part B Eng 2015;68:170–75.

[60] Lee H-F, Yu HH. Study of electroactive shape memory polyurethane–carbon nanotube hybrids. Soft Matter 2011;7(8):3801–07.

[61] Sahoo NG, Jung YC, Goo NS, Cho JW. Conducting shape memory polyurethane-polypyrrole composites for an electroactive actuator. Macromol Mater Eng 2005;290(11):1049–55.

[62] Schmidt AM. Electromagnetic activation of shape memory polymer networks containing magnetic nanoparticles. Macromol Rapid Commun 2006;27(14):1168–72.

[63] Wang W, Liu Y, Leng J. Recent developments in shape memory polymer nanocomposites: actuation methods and mechanisms. Coord Chem Rev 2016;320–321:38–52.

[64] Buckley PR, McKinley GH, Wilson TS, Small W, Benett WJ, Bearinger JP, et al. Inductively heated shape memory polymer for the magnetic actuation of medical devices. IEEE Trans Biomed Eng 2006;53(10):2075–83.

[65] Narendra Kumar U, Kratz K, Behl M, Lendlein A. Shape-memory properties of magnetically active triple-shape nanocomposites based on a grafted polymer network with two crystallizable switching segments. Express Polym Lett 2012;6(1):26–40.

[66] Yu X, Zhou S, Zheng X, Guo T, Xiao Y, Song B. A biodegradable shape-memory nanocomposite with excellent magnetism sensitivity. Nanotechnology 2009;20(23):235702–11.

[67] Mohr R, Kratz K, Weigel T, Lucka-Gabor M, Moneke M, Lendlein A. Initiation of shape-memory effect by inductive heating of magnetic nanoparticles in thermoplastic polymers. Proc Natl Acad Sci 2006;103(10):3540–45.

3 Characterization techniques for shape memory polymers

3.1 Introduction

Shape memory polymers (SMPs) are characterized by different characterization techniques. An in-depth analysis of SMP is necessary to understand the relationship between structure and shape memory effect (SME) of SMP. Extensive chemical and structural characterization is needed to investigate the nature of SMP. Appropriate and efficient structural characterizations of SMP offer the ideal application of SMP. The structure of SMP is studied by nuclear magnetic resonance spectroscopy and Fourier-transform infrared spectroscopy. The investigation of morphology of SMP is essential to understand the SME. The morphology of SMP is investigated by using microscopic techniques such as scanning electron microscopy (SEM), transmission electron microscopy (TEM) and atomic force microscopy (AFM), and scattering techniques such as wide-angle X-ray diffraction (WAXD) and small-angle X-ray diffraction (SAXD). The thermal transition temperature, i.e., T_{trans}, which is crucial for SMPs is characterized by differential scanning calorimetry (DSC) and dynamic mechanical analysis (DMA). The quantification of SME is a key characterization for potential applications of SMP. The cyclic, thermomechanical tensile test is widely used technique for quantification of shape memory properties. SME is not an intrinsic property of the SMP; it depends on the molecular structure, processing and programming conditions. The suitable multiple characterizations will give the detailed understanding of SME of SMP. The various properties of SMP characterized by different techniques are presented in Table 3.1. The different techniques for characterization of chemical structure, morphology, thermal properties and quantification of SME are discussed below.

3.2 Characterization techniques of SMP

3.2.1 Nuclear magnetic resonance spectroscopy

Nuclear magnetic resonance (NMR) spectroscopy is carried out to evaluate the structure of the SMP. Usually proton (^{1}H) and carbon (^{13}C) NMR spectroscopic technique is used for this purpose. This technique is used to determine the molecular weight, degree of branching, repeating units and distribution, tacticity and the presence of the nonreacted monomers [1]. This spectroscopic technique can also be performed to investigate the kinetics of the polymerization reaction, cross-linking reaction, composition of copolymer, degradation nature and so on. Furthermore, this technique can be used to understand the effect on the nanomaterials within the polymer matrix.

https://doi.org/10.1515/9783110570175-003

Table 3.1: Techniques for characterization of SMP.

Properties	FTIR	XRD	NMR	Raman	DSC	DMA	TEM
Functional groups	++	–	++	+	–	–	–
Chemical composition	++	–	++	+	–	–	–
Degree of crystallinity	–	++	+	+	++	+	–
Chain conformation	+	+	++	+	–	–	–
Molecular weight	–	–	+	–	–	–	–
End groups	++	–	++	+	–	–	–
T_g and T_m	–	–	–	–	++	++	–
Viscosity	–	–	–	–	–	+	–
Kinetic of reaction	++	–	++	+	++	–	–
Modulus	–	–	–	–	–	++	–
Morphology	+	++	+	+	+	+	++
Compatibility	+	+	+	+	++	++	++
Shape recovery behavior	–	–	–	–	–	+	–

++, strong technique; +, weak technique; –, inadequate technique; FTIR, Fourier transformed infrared spectroscopy; XRD, X-ray diffraction; NMR, nuclear magnetic resonance spectroscopy; DSC, differential scanning calorimetry; DMA, dynamic mechanical analysis; TEM, transmission electron microscopy.

NMR study of polymers is carried out in solid state or in swollen state. Solid-state NMR is known to be a very effective tool to obtain detailed information of the phase, structural and dynamic properties of amorphous and semicrystalline polymers, both in their pristine and in blend or composite systems [2]. This is because the spectroscopic properties such as chemical shifts, relaxation times and line widths depend on the local structure and dynamics of molecules. It can also provide the information at nanometer and molecular level associated with the SME as well as the state of programming SMP. The SME could be observed by a solid-state ^{1}H double quantum NMR experiment of the polymer network. In general, the position of the maximum of double quantum excitation shows the strength of the dipolar coupling. The shorter the excitation time at the maximum of double quantum coherence shows the higher is the dipolar coupling. The morphology of polymers can be studied by proton spin diffusion methods because the rates of spin diffusion are sensitive to domain sizes on the order of 1–20 nm.

The SME of oligo [(L-lactide-ran-glycolide)] dimethacrylates through solid-state NMR was investigated by Bertmer et al. [3]. They reported that below T_g the unstretched (original shape) sample because of the rigidity of the polymer network showed a maximum of double quantum at an excitation time of 20 µs. However, the stretched sample showed a maximum of double quantum at an excitation time of 18 µs. After stretching the sample, the chain length between the cross-linking points is elongated and partially oriented along the stretching direction. So the dipolar coupling is increased which led to a shorter excitation time than the unstretched sample. Moreover, the normalized signal intensity of the stretched sample is much higher

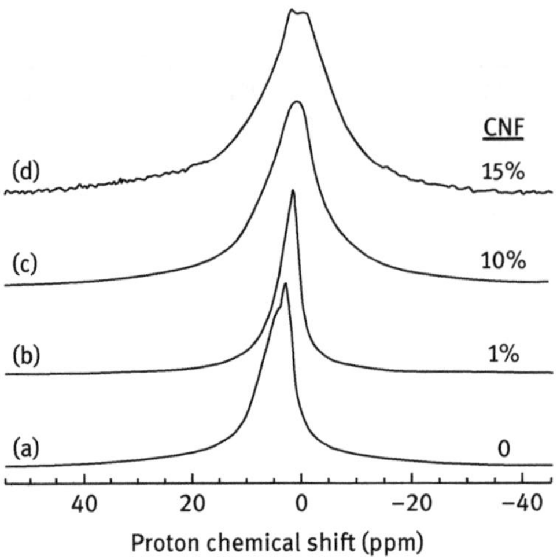

Figure 3.1: ^{1}H NMR spectra of shape memory thermoplastic polyurethane and its nanocomposites.

than the unstretched sample. This fact indicates that the efficiency of excitation of double quantum coherence is strongly increased due to the higher dipolar coupling. After heating, the temporary shape above T_g recovered the permanent shape. A double quantum measurement at room temperature exhibits an almost identical buildup curve similar to the unstretched sample. Thus, the SME is reversible and can be confirmed from solid-state NMR study. Powers et al. [4] studied the structure and dynamics of a shape memory thermoplastic polyurethane/carbon nanofiber (CNF) nanocomposite by solid-state NMR. The nanocomposites exhibited shifting and broadening of peak as compared to the neat polymer (Figure 3.1). The change in chemical shift could be ascribed to ring current shifts from CNFs in close proximity to the soft segments or to magnetic susceptibility differences in the composites. The change in line width could be explained either to a restriction of chain mobility or magnetic susceptibility variations introduced with the CNFs. The proton NMR spectra of stretched samples of both the neat polymer and nanocomposite exhibited the downfield chemical shift and broadening of the peak. However, the unstretched polyurethane and its nanocomposite samples showed similar chemical shift and broadening of peak. This result reflects that the local magnetic environment for the amorphous chains is identical in neat polymer and its nanocomposites, indicating that these chains are no longer in close proximity to CNFs. However, highly stretched samples showed similar chemical shifts for with and without CNFs imply that the mobile amorphous phase is no longer in close proximity to CNFs. These suggested that nucleation is promoted at the surface of CNFs, thereby separating the mobile phase from CNFs. The morphology of stretched and as-cast films was studied by the spin diffusion techniques to characterize the mobile and rigid domain sizes. The spin diffusion pulse sequence builds a polarization gradient between the mobile and rigid domains, and the domain sizes can be calculated by monitoring the rate of magnetization exchange between them.

The author reported that highly stretched sample led to both smaller domain sizes and smaller spacings than the low stretched sample.

The presence of hydrogen bonding in the shape memory polyurethane has been observed by the ^{13}C NMR analysis, which is important factor for the SME of polyurethanes [5]. In ^{13}C NMR spectra, carbon atoms sensitive to hydrogen bonds are quaternary aromatic carbons from 4,4′-methylenebis(phenyl isocyanate) (MDI) units adjacent to urethane moiety. A shifting of the signal at $\delta = 120.4$ downfield to $\delta = 150.9$ corresponding to aromatic –NH-C<– groups is observed as a result of a decrease in electron density owing to the participation of the neighboring –NH group in the hydrogen bond. In the solid-state NMR spectrum only one signal of the –NH-C<– group was observed, indicating that all the NH groups form hydrogen bonds of the same strength, either with carbonyl oxygen atoms of another urethane moiety or ether oxygen atoms of the soft segments. Kalita and Karak [6] evaluated the structure and the degree of branching of shape memory hyperbranched polyurethane by the proton NMR. Proton NMR spectrum of polyurethane resulted by the presence of urethane linkages, toluene diisocyanate (TDI) moieties and other characteristic groups such as double bond, chain $-CH_2-$ and terminal $-CH_3$. The peaks at $\partial = 0.80$ ppm, $\partial = 1.19$ ppm and $\partial = 1.47$ ppm are due to the terminal methyl group, all terminal $-CH_2-$ groups and the protons for $-CH_2-$ groups attached next to the terminal methyl group of the fatty acid chain of the monoglyceride of the

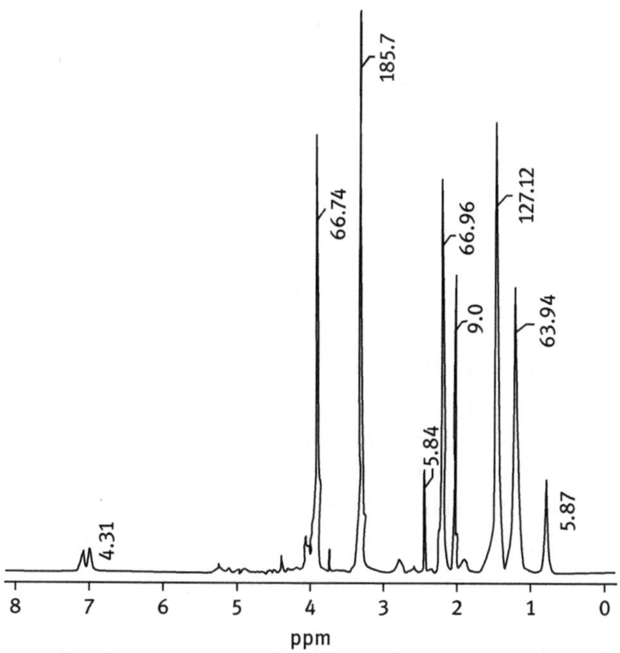

Figure 3.2: ^1H NMR spectrum of shape memory hyperbranched polyurethane.

oil, respectively (Figure 3.2). The protons of allylic $-CH_2-$, $-CH_2-$ adjacent to $-O-$ of urethane group and $-CH_3$ of TDI showed peaks at ∂ = 2.19 ppm, ∂ = 2.23 ppm and ∂ = 2.68 ppm, respectively. The $-CH_2-$ protons of triethanolamine moiety attached to the urethane linkages and $-CH_2-$ protons adjacent to –OH groups were found at ∂ = 3.21 ppm and ∂ = 3.92 ppm, respectively. The observed integration ratio of these two peaks was 2.78, which indicated that out of three –OH groups of triethanolamine moiety, 2.78 groups were substituted, i.e., the degree of branching is about 0.93. The observed degree of branching indicated that the material exhibited highly branched structure. The hyperbranched polymer exhibited higher shape recovery behavior than the linear counterpart.

3.2.2 Fourier-transform infrared spectroscopy

Fourier-transform infrared (FTIR) spectroscopy is a commonly used technique for characterizing the SMP. This technique probes the vibrations or rotations of the atoms of a molecule [1]. This spectroscopy records the absorbance or transmittance intensity in percentage as a function of wave number (cm^{-1}) or wavelength (nm) of the materials under investigation. The rotational and vibrational energies of any specific group or atom that fall in the infrared region results in the characteristic absorption band. These spectra serve as fingerprints for particular types of bonding, allowing for their identification and relative quantification. Thus this spectroscopy is mostly used to determine the functional groups present in the structure of SMP [7]. This technique is also used to identify a polymer, observing the polymerization stage in real time, and to characterize the structural changes in different reaction conditions. The formation of SMP nanocomposites can also be characterized by this technique. In addition, morphology of the SMPs that have an important role on SME can be investigated by this technique. It can be identified whether the functional groups present in the SMP facilitated the SME or are to be used for subsequent modification of the polymer to reach new SMP.

The structure of the shape memory polyurethane has been studied by this technique. The formation of urethane linkage can be confirmed by the shifting in the peak positions of hydrogen-bonded N–H and C=O groups. The characteristic bands of FTIR spectra of shape memory polyurethanes synthesized from TDI, polycaprolactone (PCL), 1,4-butanediol (BD) and triethanolamine are given in Table 3.2 (Figure 3.3) [6]. The completion of the polymerization reaction can be confirmed from the FTIR spectrum. The absence of NCO band confirmed the completion of reaction. The morphology of the segmented polyurethane can be observed by studying the bands of the hydrogen-bonded ordered groups that have some of regularity and hydrogen-bonded disordered groups in the region that are amorphous. The band of ordered hydrogen-bonded carbonyl groups is sharper and is observed at a lower frequency than the corresponding band observed in disordered hydrogen-bonded

Table 3.2: The bands of FTIR spectrum of shape memory polyurethane.

Band position (cm^{-1})	Functional groups
3,405–3,435	–NH stretching vibrations
2,850–2,945	$-CH_2$ symmetric and anti-symmetric stretching vibrations
1,630–1,745	Amide I, >C=O stretching vibrations
1,450–1,460	$-CH_2$ scissoring, $-CH_3$ deformation and $-CH_2$ bending vibration
1,470–1,540	Amide II
1,245–1,255	Amide III, ν C–N and in plane –N–H deformation
1,045–1,065	–O–C=O stretching vibration of urethane/ester group
870	>C–O stretching and $-CH_2$ rocking vibration
711–720	Amide IV

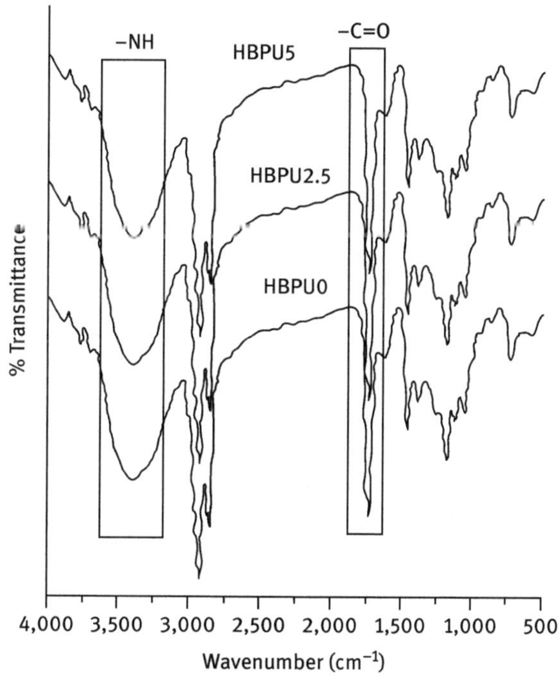

Figure 3.3: FTIR spectra of shape memory polyurethane.

samples (Figure 3.4) [8]. This is a consequence of ordered structure, where the chains are closer together and the hydrogen bonds are stronger. The degree of ordering in the hard segment (HS) can be determined by the FTIR analysis. The urethane carbonyl groups located in the small HS microdomains solubilized in soft segment phase are likely nonhydrogen bonded whereas those located at the interfacial zone between hard domains and soft matrix could be hydrogen bonded or free, depending on the configuration of these groups relative to the nearby NH proton donors. The relative absorbance of the two urethane carbonyl bands can be termed as an index of the

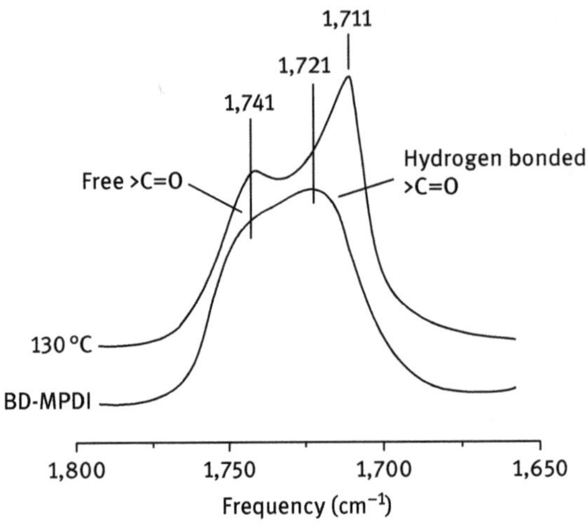

Figure 3.4: Carbonyl stretching region of the BD-MPDI polyurethane, bottom as-cast film; top film held at 130 °C for 20 min.

degree to which this group participates in hydrogen bonding. The fraction of hydrogen bonding may be expressed as [9]

$$X_B = \frac{C_B}{C_F + C_B} = \frac{\frac{A_B}{\varepsilon_B}}{\frac{A_F}{\varepsilon_F} + \frac{A_B}{\varepsilon_B}}$$

where A, C and ε are the absorbance, concentration, and extinction coefficient of bonded (B) and free (F) carbonyl groups, respectively.

The fraction of hydrogen bonding has been observed for the polyether-based polyurethane. The fraction of hydrogen-bonded urethane carbonyls increases with increase in the amount of HS content, indicating an increase in hard domain ordering. In addition, the degree of ordering of the HS can be obtained from hydrogen-bonding index (α), which can be determined from the area under the peak of free carbonyl groups (A_{FCO}) and hydrogen-bonded carbonyl groups (A_{HCO}) [10]:

$$\alpha = \frac{A_{HCO}}{A_{FCO} + A_{HCO}}$$

The effect of various diisocyanates on microphase-separated structure of polyurethanes has been studied by FTIR analysis, monitoring the change in the relative intensities of free and hydrogen-bonded carbonyl (C=O) peaks [11]. The carbonyl peak for the urethane based on aromatic and symmetrical 1,4-phenylene diisocyanate (PPDI) showed at 1,695 cm^{-1}, while the carbonyl peak for a homologous urethane based on aliphatic and symmetrical 1,4-cyclohexyl diisocyanate (CHDI) showed at a slightly lower wave number of 1,683 cm^{-1}. The urethane compound based on aliphatic but unsymmetrical bis(4-isocyanatocyclohexyl) methane (HMDI) showed a fairly sharp peak at 1,685 cm^{-1}, which also has a distinct shoulder at 1,695 cm^{-1}. The appearance of shoulder is most probably owing to the presence of several isomers (*trans*/*trans*,

trans/*cis* and *cis*/*cis*) in HMDI. The isomers insist the packing in the solid state and in turn the crystal structure of urethane, as a result prevents the formation of the well-ordered hydrogen-bonded structure. The sharp peak at 1,685 cm^{-1} is a representative of well-ordered hydrogen-bonded structure, whereas the shoulder at 1,695 cm^{-1} is due to inadequate hydrogen bonding. It was concluded based on the hydrogen-bonded carbonyl peaks that PPDI- and CHDI-based polyurethanes exhibited self-organization and formation of well-ordered HSs, on the other hand no change in the carbonyl region or no phase separation was observed for MDI- and HMDI-based polyurethanes.

3.2.3 X-ray diffraction

X-ray diffraction (XRD) is performed to evaluate the degree of crystallinity, amorphousness, crystal structure and size of crystals of the SMP. SMP nanocomposites can also be characterized by this technique. Two types of X-ray diffraction such as WAXD and SAXD are used to characterize the polymer. WAXD gives the information regarding the degree of crystallinity, orientation of crystalline region and nature of ordering structure [12]. On the other hand, SAXD is utilized to obtain information on the dimension of small crystalline regions as lamellae and the presence of voids and their shapes [13]. The combined study of WAXD and SAXD provides the quantitative characterization of nanostructure and crystalline structure of polymer [14]. The investigation on influence of degree of crystallinity and strain-induced crystallinity on shape fixity and shape recovery behaviors is essential for SMP. The morphological features and microphase-separated structure of SMP have strong influence on the SME, which can be analyzed by the SAXD. Moreover, careful applications of this technique such as interdomain spacing, domain boundary diffuseness and the degree of microphase separation can be measured. SAXD provides a number of significant advantages and although interpretation is less direct than are the TEM or AFM techniques. The sample preparation for SAXD avoids the potential for staining artifacts that may take place with TEM analysis. X-rays are less destructive to the samples than are the electron beams used in TEM analysis. SAXD is employed to characterize the structure of domain in the size range of tens of nanometers. X-rays are scattered by the electrons of a sample, and the amplitude of the scattering radiation is provided by the three-dimensional Fourier transformation of the electron density distribution. The information concerning morphology of the material is investigated by analyzing the intensity as a function of angle. The position of the maximum SAXD peak in the respective scattering intensity profile is used to determine the Bragg or d spacing in the material, which may be interpreted as an average interdomain spacing occurring from the presence of ordered structure.

SAXD profiles for patterns of phase-separated shape memory polyurethanes show a distinct scattering peak. This implies a quasiperiodic fluctuation in electron density within these materials. The effect of symmetric isocyanate (MDI) and asymmetric isocyanate (2,4-TDI) on the morphology of segmented polyurethanes

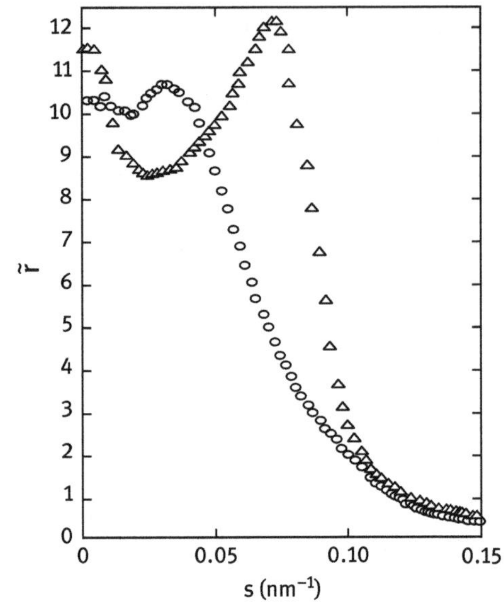

Figure 3.5: SAXD profiles of shape memory polyurethanes for MDI/BD (open circles) and TDI/EG (open triangles).

has been studied by SAXD analysis [15]. Both the polyurethanes show curves with a single maximum followed by a gradual decrease in intensity. The peak is better defined and appears at larger scattering angle for the asymmetric TDI/EG (ethylene glycol)-based polyurethane (Figure 3.5). The decrease in intensity is also steeper for this polyurethane. The broad maxima scattering pattern of symmetric MDI/BD-based polyurethane indicates long-range heterogeneities in this material while the sharp maxima pattern of asymmetric TDI/EG-based polyurethane suggests a narrow distribution in interdomain spacing. The interdomain spacing is calculated from Bragg's law and found to be 12.5 and 22 nm for the asymmetric TDI/EG-based polyurethane and symmetric MDI/BD-based one, respectively. The larger interdomain spacing indicates that the MDI/BD HSs are partially ordered into a paracrystalline structure and as a result the MDI/BD domain is thicker. The smaller interdomain spacing in TDI/EG-based polyurethane indicates the absence of strong preferential ordering of TDI/EG HSs. The morphology of poly(tetramethylene glycol, M_n = 1,000 or 2,000 g/mol) (PTMO)-based polyurea has been investigated by SAXD [16]. The arising of a first-order scattering peak in their respective intensity profiles indicates the presence of a microphase-separated morphology in polyureas at ambient conditions. The average interdomain spacing is found to increase with the increase of molecular weight of the soft segment (PTMO). The microphase separation structure of PTMO and PPO-PEO soft segment-based polyurethanes has been observed by the SAXD analysis [17]. Both the scattering patterns of polyurethanes showed a scattering peak, which implied the presence of a two-phase structure. The scattering intensity of the PTMO-based one was higher than that of the poly(propylene oxide)-poly(ethylene oxide) (PPO-PEO) one, indicating that PTMO-based polyurethane exhibited higher degree of phase-separated structure. PTMO soft

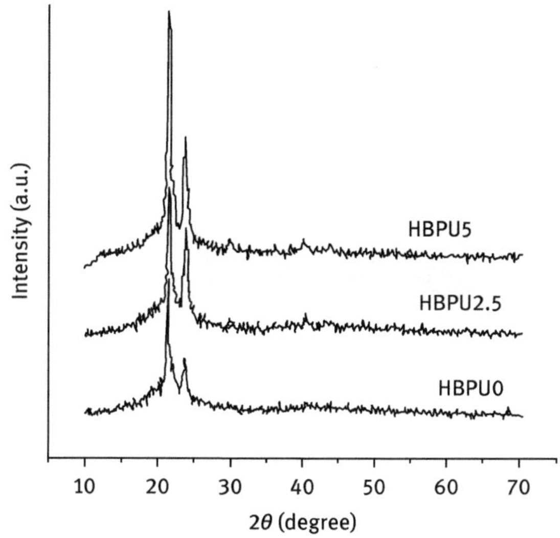

Figure 3.6: XRD diffractograms of shape memory polyurethanes.

segment chains have no side groups and are more flexible than PPO and thus kinetic barrier for PTMO is smaller than that for PPO. As a result, MDI/BD HSs could be more readily excluded from the PTMO soft segment matrix.

Kalita and Karak [6] studied the effect of amount of multifunctional moiety on the shape memory behaviors of biobased polyurethanes. There are two strong diffraction peaks at 2θ = 21.98 and 23.88° due to the (100) and (200) planes of PCL crystals present in the shape memory polyurethane. They reported that crystallinity was found to increase with the increase in the amount of multifunctional moiety content (0–5 wt%) (Figure 3.6). This is attributed to the increased secondary interactions in the polymer as observed by the FTIR analysis. Due to the increased secondary interactions, the HSs were forced to stay close to the cross-linking points and the relative order of HS alignment increased the intensity of the peak. The materials exhibited excellent shape fixity and shape recovery because of the increased crystallinity of the systems.

The mutual study of SAXD and WAXD facilitating the in-depth observation of crystallization, domain sizes, interdomain spacing and degree of phase separation offers superior insights on structural processes during programming and recovery of SMPs. By correlating the chemical structure of SMP with the parameter such as crystal sizes, domain sizes and orientation investigated from X-ray scattering in the programming of shape memorization confer well understanding to develop new SMP.

3.2.4 Scanning electron microscopy

The scanning electron microscope is a type of electron microscope that creates magnified images by the electrons emitted when the primary electrons coming from the

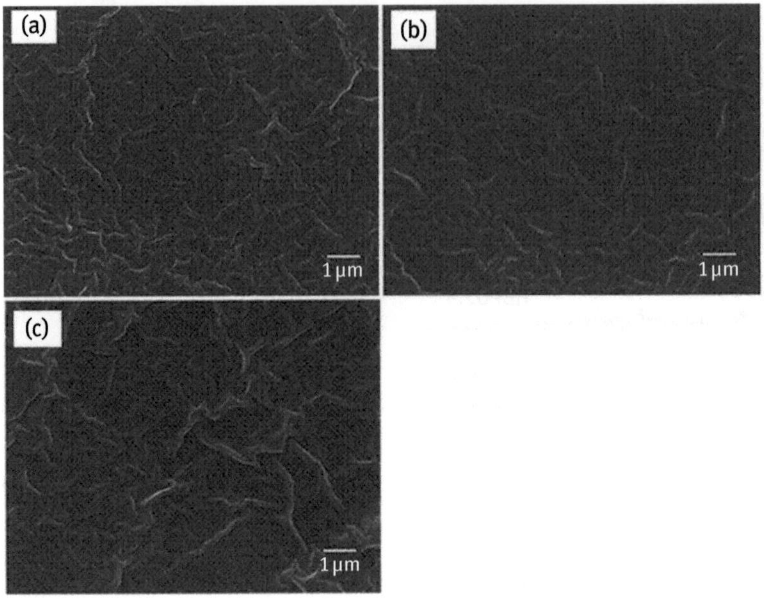

Figure 3.7: SEM micrographs of shape memory hyperbranched polyurethane: (a) HBPU0, (b) HBPU 2.5 and (c) HBPU 5.

source strike the surface and are inelastically scattered by atoms in the sample [18]. The morphological analysis of the SMP and their nanocomposites is done by employing the SEM technique [1]. SEM is usually used to study surface topography of polymer blends, block and graft copolymers and nanocomposites. Moreover, it can also be utilized to obtain information of crystallographic orientation and composition of polymeric materials by the help of energy-dispersive spectroscopic attachment. The phase-separated morphology of the SMP can be analyzed by this technique regarding domain sizes and number of phases. A comprehensive study on structure and distribution of domains is important to understand the SME of SMP. Kalita and Karak [6] studied the morphology of hyperbranched shape memory polyurethanes by SEM. They observed two-phase morphology, i.e., bright and dark regions corresponding to hard and soft segments of the polyurethanes, respectively (Figure 3.7). It was also observed that HSs that were homogeneously distributed in the soft segment matrix resulted in the good SME.

3.2.5 Transmission electron microscopy

TEM is used to determine the distribution of different phases, internal structure, defects structure and so on present in SMPs and their nanocomposites [1, 19]. It can also be employed to obtain the information about crystallography and chemical composition of the SMP. Under favorable conditions, this technique can be determines at the 0.1 nm level, but such high-resolution examination is hardly possible with polymers. In TEM, the electron beam is allowed to penetrate to the very thin film and the transmitted electrons are directly used to generate the image of the specimen with the help of lenses. Varying the electron density in regions with different morphology leads to a contrasted image. The phase-separated structure of a block copolymer or a blend SMP can be studied by TEM analyses. The characteristic microstructural features of SMP such as spatial variations in density, crystallinity, crystal orientation and distribution, dimension of crystalline and amorphous regions are often important to characterize. Light scattering techniques are potential tools to characterize these morphological features. However, real-space visual characterization of the domain dimensions, shape, orientation and dispersion of microdomains are not viable by this technique. By using TEM technique the image of nanoscale morphologies of polymer can be observed, even though it is a challenging task. Polymers consist mostly of low atomic number elements, whose elastic interactions with high-energy electrons are moderately weak. There is almost no absorption of electrons directed on samples as thin enough for TEM. However, this technique has the ability to image at high magnifications, provided there is sufficient contrast between the electron density of the two phases. Thus, contrast-enhancing techniques such as staining were required [20]. This method introduces heavy metals into particular reactive sites via a chemical reaction or by absorption. Ruthenium tetroxide (RuO_4) is used selectively for compounds containing double bonds to enhance the contrast.

Phase-separated structure of shape memory polyurethanes could be observed by TEM analysis as the dimensions of the domains are in the order of 5–100 nm. Serrano et al. [21] reported the microphase-separated structure observed in TEM images of TDI/PBU/BDO polyurethanes even without staining, in samples with 55% and 75 wt% HSs. They observed ordered lamellar structure of HS with the domain size of the order of 100 Å. Koutsky et al. [22] reported the application of TEM for the domain morphology study of polyether and polyester urethanes having 38 wt% HSs. They used solvent etching and iodine staining of the samples. In both the urethanes, dark domain, i.e., HS about 5–10 nm wide was observed (Figure 3.8). Li and coworkers [23] have investigated the morphology of polybutadiene-based polyurethanes having different HS (MDI/BD) concentration using high-voltage electron microscopy. Osmium tetroxide was used to selectively stain chemically unsaturated moieties. A rod-like or lamellar structure was observed in polyurethanes when the HS content varied from 0.42 to 0.67. The HS phase is dispersed in a matrix of soft segments, in the form of either short cylinders or spheroids when the HS content is less than 31 wt%. At very high HS contents, isolated polybutadiene soft segment microdomains are embedded in a HS matrix.

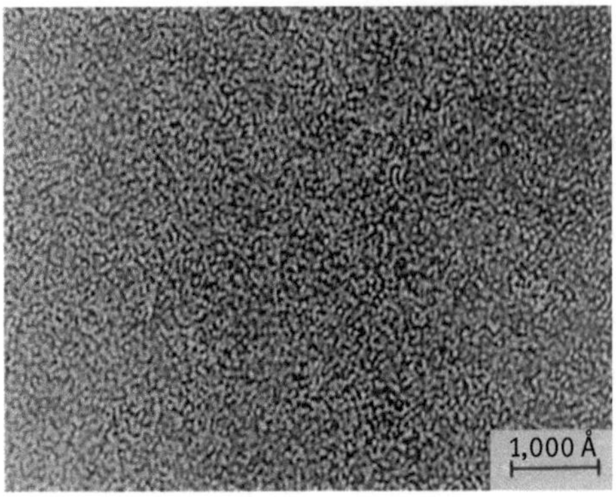

Figure 3.8: TEM image of polyester urethane which was solvent etched and stained by iodine.

3.2.6 Atomic force microscopy

The surface morphology and phase segregation behaviors of SMP at nanoscale levels can be carried out by AFM. Conformational and chain order, crystalline order, lamellar structures and lamellar surfaces can also be investigated by AFM [24]. This basically consists of a sharp tip or probe mounted to a cantilever. A deflection of the cantilever arises when the tip is brought in close proximity to a surface, as a result of attractive or repulsive forces of interaction between a tip and the surface. This deflection is detected through laser beam by reflecting an incident beam off the flat top of the cantilever. The changes in direction of reflected beam due to the cantilever deflection are recorded by the position-sensitive photodiode. Usually both the AFM and TEM are well-established techniques for visual representation of the morphology of polymers. However, TEM experiments are also limited by the possibility of beam damage, and are also tedious and time consuming because of the microtomy involved in samples preparation into few tens of nanometer thin sections. Imaging at high magnifications under TEM can also lead to misleading phase-contrast artifacts observed at a scale length of approximately 100 Å under slight defocus conditions. The topography, the degree of phase segregation and local stiffness can be measured by means of AFM [25]. A comprehensive relationship between nanoscale morphology of SMP and their SME can be understood by means of AFM analysis.

The topographical images of hyperbranched shape memory polyurethanes prepared from 4,4′-diphenylmethane diisocyanate, poly(butylene adipate) glycol and hyperbranched polyester as chain extender have been investigated by tapping-mode AFM [26]. Two regions were observed, i.e. bright and dark, the brighter region indicated the isocyanate-rich dispersed hard phase and the darker area indicated the polyol-rich dispersed soft phase. The phase separation is obtained in the shape memory polyurethane due to the incompatibility of hard and soft domains. The difference in mechanical properties of the soft and hard domains leads to a force differences in AFM measurement and hence different image of morphology is found.

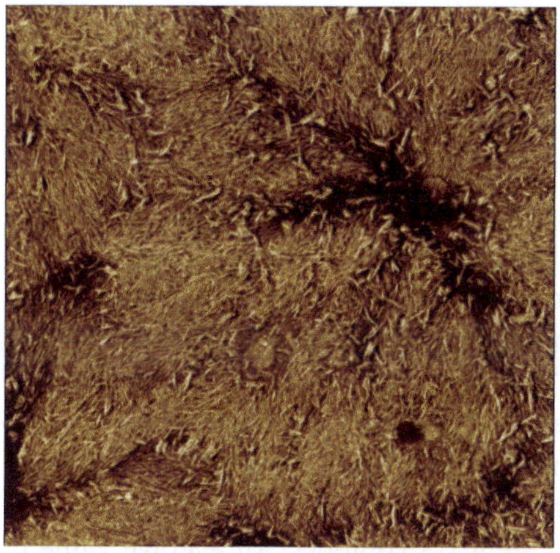

Figure 3.9: Microphase-separated morphology and spherulitic structure of the polymer observed in AFM.

The microphase separation influences the SME, where hard domain dispersed in the soft domain acts as filler lead to maintain the original shape of the SMP. The nanoscale morphology of segmented polyurethane synthesized from poly(tetramethylene oxide)) as soft segments, and BD extended piperazine as HS has been elucidated by AFM analysis [27]. A spherulitic morphology with spherulite size of 2 μm was observed (Figure 3.9). It was noticed that hard domains were not isolated from each other but have some extent of connectivity. The hard domain connectivity was found to increase with the increase of HS content. The length of HS was observed approximately 60 Å. It was suggested that the HSs within the microdomains were oriented preferentially along the tangential direction of the spherulite. The microphase-separated nanoscale morphology of polyurethanes with varying NCO/OH ratio prepared from MDI, BD and poly(tetrahydrofuran) polyether polyol has been observed by AFM [28]. The sample with NCO/OH ratio of 0.940 exhibited a wavy morphology of the HSs that assembled into larger domains with area of few 100 nm^2 while the sample with NCO/OH ratio of 1.025 showed sharp features of straight whiskers with random orientation of the HSs. The whiskers with a narrow apparent full width of about 20 nm vary in lengths from 30 to 200 nm. However, the sample with NCO/OH ratio of 1.150 showed a fiber bundle-like morphology of the HSs. The materials exhibited differently significant morphology with varying molar ratio.

3.2.7 Raman spectroscopy

Raman spectroscopy relies on the inelastic scattering of monochromatic light usually from a laser in the visible, near-infrared or near-ultraviolet range with materials. This is a potential technique for analyzing molecules without a permanent dipole moment. Raman spectroscopy is highly sensitive to the material crystallinity, orientation and temperature [29]. Raman spectroscopy can be utilized to monitor the polymerization process and can provide information on the extent of polymerization and structural information on the end-product [30]. It can also be used to study the kinetics of the polymerization reaction and degradation processes of polymer in molecular level. In addition, this technique is used to study of stress transfer between nanomaterial and the matrix polymer. The microphase separation and orientation of domain in SMPs relating to the SME can be studied by Raman spectroscopy using polarized light.

Polarized Raman spectroscopy was performed to study the phase separation and molecular orientation for shape memory polyurethanes with various HS contents [31]. In the Raman spectrum a highest peak at 1,616 cm^{-1} vibrational mode was observed and the magnitude of the peak was studied with various HS contents. The orientation of the crystalline HS was studied from the obtained in-plane vibrational mode of the two benzene rings at 1,616 cm^{-1}. The HS orientation was obtained from the depolarization ratio (ρ) with respect to different HS contents. An average value of $\rho = 0.71$ suggests that the benzene rings of HS are initially randomly oriented in the materials at low HS content. As the HS content is above a particular value of 25%, there is sudden increase in the polarized signals. This indicated that a new phase of HS is decoupled from the soft matrix due to the high incompatibility between the hard and soft segments. The depolarization ratio is 0.45 (HS content between 25% and 45%), indicating that the HS domains orient themselves in a certain configuration apart from a random orientation. A value of $\rho = 0.35$ implies that the benzene rings are now to a large extent perpendicular to the film surface at high HS content (>45%). It could be concluded that the observed ρ values from 0.71 to 0.35 conferred the presence of a transition region in which the benzene rings in the HS start to demix from the matrix and orient themselves to lie perpendicularly to the film surface. The kinetics of polymerization of thermoplastic polyurethane reacted from MDI, BD and hydroxyl-terminated poly(butylene adipate) has been studied by Raman spectroscopy [32]. The different bands of Raman spectra of polyurethane are given in Table 3.3. The progress of the polymerization could be monitored by observing the corresponding band intensity. The band at 1,530 cm^{-1} is similar to the isocyanate asymmetric and symmetric stretching vibrations and its intensity decreases with increasing polymerization time and diminishing the peak intensity confirms completion of the reaction.

Table 3.3: Bands of Raman spectra of shape memory polyurethane.

Raman shift (cm^{-1})	Assignment
2,275	N=C=O (asymmetric stretching)
1,732	Ester C=O stretching, urethane amide I (C=O stretching)
1,612	C=C stretching (aromatic ring)
1,530	Phenylene ring stretching, urethane amide II (C–N and N–H stretching)
1,445	N=C=O (symmetric stretching)
1,303	Urethane amide III
920–930	C–H bending (aromatic ring)

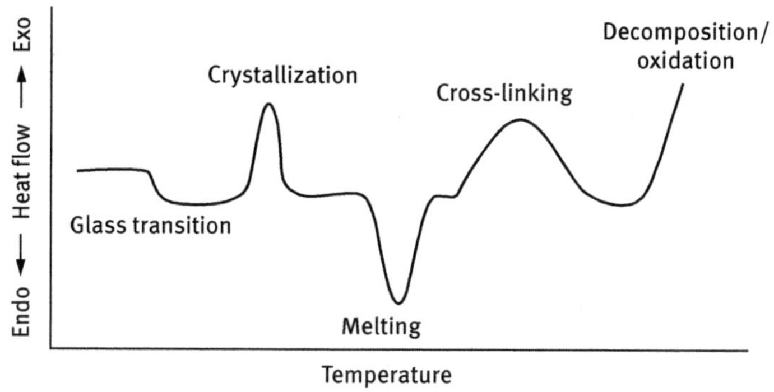

Figure 3.10: A typical DSC curve of heat flow as a function of temperature.

3.2.8 Differential scanning calorimetry

DSC is a technique to measure the change of difference in the heat flow rate to the sample as a function of temperature. A typical DSC curve of heat flow as a function of temperature is shown in Figure 3.10. DSC is a very potential tool to evaluate the thermal properties of the SMP. Thermal properties are key characteristics of SMPs, especially switching/transition temperature (T_{trans}), i.e., either glass transition temperature or melting temperature. The glass transition temperature (T_g), melting temperature (T_m), degree of crystallinity, crystallization time and temperature, degree of polymerization, kinetics of reaction, oxidation/decomposition, amount of endothermic or exothermic energy can be measured by means of DSC analysis [1, 33]. T_g is defined as the temperature region at which amorphous phase of polymer is transformed from the glass state to the viscoelastic state. This second order endothermic transition observes as a step transition. An endothermic peak is observed for the melting transition, i.e., first-order transition. However, T_g and sub-T_g transitions are usually too weak or broad to be detected by DSC and can be better identified by DMA. In DSC experiments typically the data of the second heating run are analyzed after the thermal history of samples as processing is eliminated through heating above the highest T_{trans} in the first heating run.

The crystallinity (χ_c) of the polymer can be determined by DSC, measuring the enthalpy change (ΔH), which is proportional to the peak area between the curve and the baseline. The χ_c is calculated by dividing the change in enthalpy of polymer sample by the change in enthalpy of a 100% crystalline analogue (ΔH°_m), i.e., $\chi_c = \Delta H_m/\Delta H^{\circ}_m$ [34]. It can reveal the mobility of molecular chains by specific thermal energy variation.

The effect of crystallinity on SME of varying amount of HSs content polyurethanes prepared from PCL as soft segment, MDI and BD as HS has been studied by DSC analysis [35]. All the samples exhibited endothermic peaks in the range of 40–50 °C, which indicated the melting of the crystallites of PCL soft segments. Polyurethanes with HS

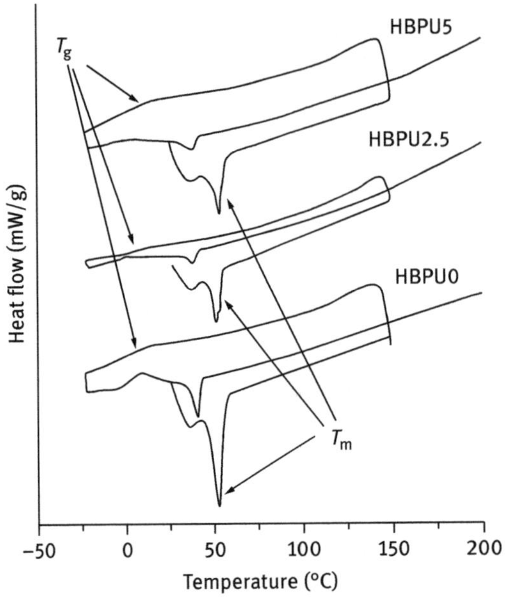

Figure 3.11: DSC curves of shape memory polyurethanes.

content is higher than 25 wt%, and another endothermic melting peak appeared at about 200 °C was attributed to the melting of HS. With the increase of HS content, the endothermic peaks representing the soft segment phase were weakened while the peaks pertaining to the HS phase were intensified, which resulted in the decrease of the crystallinity of the soft segment phase with increase of HS content. This was also in agreement with the results observed in WAXD analysis. The shape fixity was decreased with the increase of HS content, which can be attributed to the decrease in the crystallinity of the soft segment. It was concluded that the crystallization of soft segments determined the shape fixing of the shape memory polyurethanes. Thermal transition behaviors of biobased hyperbranched shape memory polyurethanes synthesized PCL as soft segment, TDI, BD, monoglyceride as HS and triethanolamine as branched generating moiety have been studied [6]. T_g and T_m were increased with the increase of branched generating moiety (Figure 3.11). This fact was due to the increased secondary interactions such as hydrogen bonding and polar–polar interaction, which make the systems more compact and rigid with the increase of branched generating moiety. The increased crystallinity enhanced the shape memory performance of the hyperbranched shape memory polyurethanes.

3.2.9 Dynamic mechanical analysis

DMA is a powerful technique to characterize the viscoelastic behavior of SMP as a function of temperature or frequency [36]. Generally, it involves applying an oscillating stress to a sample and measuring the material's response to that stress. The

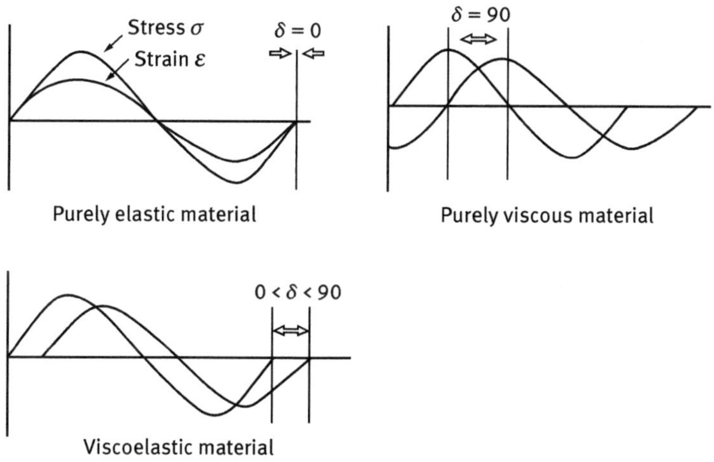

Figure 3.12: Material responses to oscillating force in DMA.

phase lag between the applied stress and measured strain reflects the material's tendency to flow (viscosity), while the material's stiffness (modulus) reflects from the sample recovery. When the material responds to the applied stress wave, an in-phase response is seen in case of perfectly elastic solid, while an out-of-phase response gives a purely viscous material (Figure 3.12) [37]. However, viscoelastic material responses fall in between these two lines. Storage modulus (E'), loss modulus (E'') and loss factor ($\tan \delta = E''/E'$) are measured by means of DMA. The storage modulus quantifies the energy stored elastically by the material upon deformation, while the loss modulus is a measure of the energy which is dissipated as heat during deformation. The storage modulus provides information regarding the stiffness of the material, while the loss factor tan δ measures the degree of molecular motion. Further, the tan δ value corresponds to the strain energy dissipated by viscous friction and higher tan δ implies the more viscous nature. The ratio of the modulus of glassy state to the modulus of rubbery state considers as an index of the extent of SME of SMPs. DMA is very sensitive to the molecular motion of the polymers and, therefore, this is a powerful tool for measuring T_g, which is an essential parameter of SMPs. The sub-T_g transitions that reflect molecular vibrations on the order of bond bending and stretching, and side group or pendant group motions can also often be seen in DMA scanning, which are too faint to be detected in the DSC or thermomechanical analyzer. An idealized scan of various DMA transitions is shown in Figure 3.13. T_g of highly cross-linked thermosets can be detected in DMA. An important characteristic of SMPs, i.e., the cross-linking density can be determined by DMA.

The effect of different HS content on the modulus of shape memory polyurethanes synthesized from PTMO, TDI and BD has been studied by DMA [38]. The glassy modulus is found to increase with the increase of HS content in the prepared shape memory polyurethanes as a result of increased physical cross-linking. The glassy

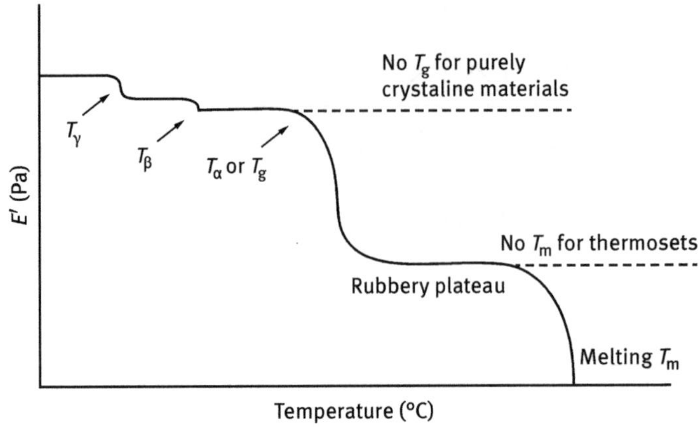

Figure 3.13: An idealized scan of various DMA transitions.

modulus (E_g) is because of the presence of elastic energy of both the PTMO and urethane phase, whereas the rubbery modulus (E_r) is because of the entropy elasticity of the two phase structures. The shape recovery is found to increase with the increase of HS content. It was revealed that higher HS content leads to higher E_g and lower E_r, resulting in a large modulus ratio, which is the cause of a better shape recovery property. The higher HS content shape memory polyurethanes exhibited higher shape fixity behavior. A large glassy state modulus results in the large shape fixity upon cooling and unloading. The tan δ value increased with the increase of HS content. Higher loss tangent value signifies that the material tends to be more viscous. Further, the tan δ value shifted to lower temperature with the increase in HS content. The effect of block length and content on the shape memory properties of segmented polyurethanes has been studied by DMA [39]. The segmented polyurethanes were synthesized from the reaction of MDI and 1,6-hexanediol as crystalline HSs and the reaction of HDI and BD as the amorphous soft segments. The rubbery plateau modulus found to increase with the increase of block lengths, which is due to the enhanced phase segregation at a longer block length could make well-developed HS domains that act as physical cross-links (Figure 3.14). The tan δ increased with the decrease of block length. As the block lengths decreased, the phase mixing found to increase whereas the hysteresis in the shape memory behavior decreased.

The miscibility and modulus of shape memory poly(vinyl chloride)/thermoplastic polyurethane blends have been investigated by means of DMA [40]. The tan δ peak of the blends found in the temperature range between those of PVC and PCL confirmed the miscibility of the blends. The blends exhibited higher rubbery plateau modulus as compared to the PVC homopolymer, which intends that HS effectively acts as physical cross-linker. Further increase in rubbery modulus was observed when the HS content increased in the polyurethanes. DMA results confirmed that PVC/polyurethane blends are phase-separated structure into HS domain and the miscible domain of PVC/PCL

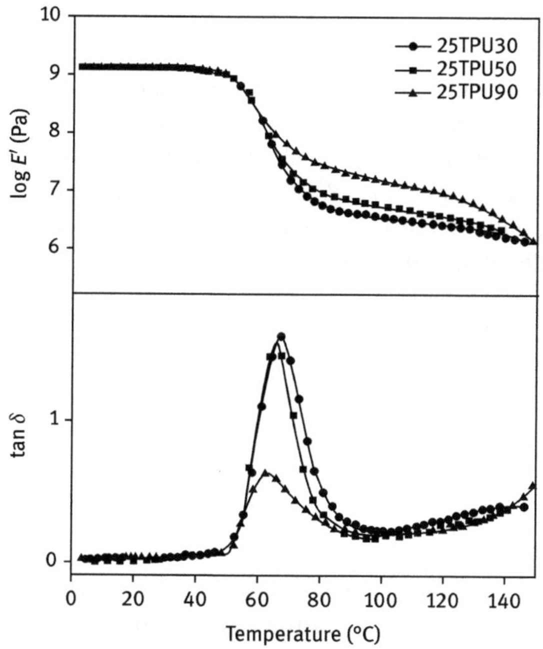

Figure 3.14: Storage modulus and tan δ of shape memory polyurethanes with different block lengths.

soft segment, which revealed the SME of blends. SME of polytetramethylene oxide/poly(acrylic acid-co-acrylonitrile) complexed gel have been investigated by DMA [41]. It was observed that molar concentrations of acrylic acid and acrylonitrile have strong influence on the modulus of the materials. The DMA profile of the complexes exhibited single major transition, while two-stage transitions were observed in higher content of acrylonitrile. This is due to the fact that only carboxylic groups of polyacrylic acid interact through H-bonding with the ether –O– atom of polytetramethylene oxide and make the complexes homogeneous amorphous phase. The materials showed the higher modulus with the higher content of acrylic acid. The modulus ratio was increased with the increase of acrylic acid content, indicating the good SME. The tan δ value of the materials found to increase with the increase of acrylic acid content, which was good agreement of the more viscous nature of the materials.

3.3 Shape memory properties

Shape memory properties of SMPs are quantified by parameters such as shape fixity, shape recovery, recovery rate and shape memory cycle life [6, 42]. These parameters are not the intrinsic properties of SMPs. They depend on the material properties and experimental conditions. The evaluation of these properties is very important to understand the SME and their potential applications. These parameters are presented in the following.

3.3.1 Shape fixity

In the quantification of SME, SMPs are heated above T_{trans} and stressed to deform it, subsequently cooling below T_{trans}. Then stress is removed and the change in strain is measured. The extent to which a temporary/deformed shape is able to fix in one cycle of shape memory test is called the shape fixity (R_f). It is similar to strain fixity and shape retention. Shape fixity depends on the structure of the SMP and the cyclic, thermomechanical tensile test conditions. It is expressed as

$$R_f\,(\%) = \varepsilon_u\,(N)/\varepsilon_m \times 100$$

where ε_m is a maximum strain and ε_u is a residual strain.

3.3.2 Shape recovery

The fixed deformed shape is reheated above T_{trans} to recover the original shape. The extent to which it recovers the original shape is called the shape recovery. Shape recovery (R_r) defines how well an original shape can be memorized by SMP. It also depends on the molecular structure of the SMP and the cyclic, thermomechanical tensile test conditions. It is expressed as

$$R_r\,(\%) = \varepsilon_m - \varepsilon_p\,(N)/\varepsilon_m - \varepsilon_p\,(N-1) \times 100$$

where ε_p is the residual strain after shape recovery.

3.3.3 Shape recovery rate

This parameter describes the speed, i.e., the rate of recovery from a fixed temporary shape to its original shape during the recovery process of SMP under the application of appropriate stimulus. It can also be said as the speed of recovery process or shape recovery speed (V_r). It is expressed as

$$V_r = \mathrm{d}R_r/\mathrm{d}T \times \mathrm{d}T/\mathrm{d}t$$

where $\mathrm{d}R/\mathrm{d}T$ is the ratio of shape recovery to temperature, and $\mathrm{d}T/\mathrm{d}t$ is the heating rate.

It is seen that V_r is a function of heating rate. It is not an intrinsic property of the SMP. It depends on the material structure and testing conditions.

3.3.4 Shape memory cycle life

The repeatability and durability of the SMP over consecutive shape memory cycles is defined as shape memory cycle life. So, the cycle life of an SMP defines the number of consecutive shape memory cycles it can sustain without failure. Here, failure indicates a noticeable decrease in the shape memory performance in terms of shape recovery and shape fixity or an actual material failure.

3.4 Characterization of shape memory properties

SME is quantified by evaluating the shape fixity ratio, shape recovery ratio and recovery rate [42, 43]. Characterization conditions have strong influenced on the shape memory properties of SMP [44, 45]. Stretching and bending techniques are conventional techniques used to evaluate the SME. The most potential and widely used technique for quantification of shape memory properties is cyclic, thermomechanical tensile test. These methods are described below.

3.4.1 Stretching technique

This is a very easy and simpler technique to quantify the SME. In this technique the specimens are heated above T_{trans} and then they are stretched to twice of their original length (L_0) and the stretched length is denoted as L_1. Immediately, the stretched specimens are put into the low temperature (much below T_{trans}) to fix the temporary shape for a specified period of time, and length is measured as L_2 after releasing the load. The cooled samples are reheated at the same elevated temperature (above T_{trans}) for the same period of time, and the length obtained is denoted as L_3. The percentage of shape recovery and shape fixity is calculated by using the following equations [46, 47]:

$$\text{Shape recovery (\%)} = [(L_1 - L_3)/L_0] \times 100$$

$$\text{Shape fixity (\%)} = [(L_2 - L_0)/L_0] \times 100$$

3.4.2 Bending technique

This is also a very simpler method to measure the shape memory properties. Various authors use this method to quantify the SME. In this technique, the specimens are heated above T_{trans} for the specified period of time and then they are folded to a ring shape. Immediately, the folded samples are put into the fixed temperature (below

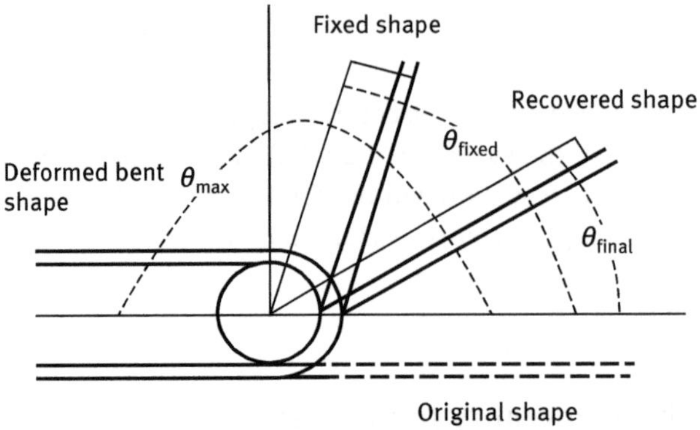

Figure 3.15: Schematic representation of bending test.

T_{trans}) for a specified period of time to fix the temporary shape. Subsequently, they are reheated at the same elevated temperature (above T_{trans}) for the same period of time for the shape recovery study. A schematic representation of bending test is shown in Figure 3.15. The percentage of shape recovery and shape fixity is calculated by using the following equations [48]:

$$\text{Shape recovery (\%)} = [(\theta_{fixed} - \theta_{final})/\theta_{fixed}] \times 100$$

$$\text{Shape fixity (\%)} = [\theta_{fixed}/\theta_{max}] \times 100$$

3.4.3 Thermomechanical cyclic tensile technique

This is a common and most widely used technique for the quantification of SME. Usually test is done in a material tensile tester equipped with constant temperature chamber. The test is performed in a series of cyclic tensile processes. The SME and recovery stress can be evaluated from the stress–strain–temperature relationship. Each cycle consists of a temporary shape which is called programming part and recovery of permanent shape which is called recovery part. The entire process is comprised of four steps: loading-cooling-unloading-recovery [42, 49, 50]. Different test procedures are adopted in the programming step. The sample is deformed under $T_{deform} < T_{trans}$ or $T_{deform} > T_{trans}$. The cooling can be carried out under stress or strain control. In the stress-controlled programming the stress is kept constant during deformation and cooling, and change in strain is measured, while in strain-controlled tests the stress on the sample is measured at defined thermal condition. Different types of recovery mode can also be carried out in the recovery step such as stress free (unconstrained recovery) and strain control (constrained recovery). In the stress-free recovery, the programmed sample is allowed to move freely as a function of temperature and the change in strain is recorded with temperature and time. The switching/transition

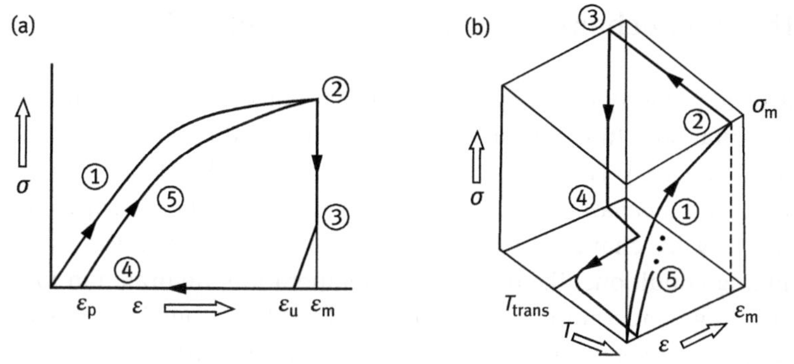

Figure 3.16: Schematic representation of the results of thermomechanical cyclic tensile test: (a) two-dimensional (ε–σ) plot and (b) three-dimensional (ε–*T*–σ) plot.

temperature (T_{trans}) can be obtained under this condition. In the strain control (fully confined) condition, the stress generated during recovery is recorded as a function of temperature and time. The maximum of recovery stress (σ_{max}) can be determined under strain-controlled condition. This stress response enables SMP to be used as actuators. Based on the application, the shape recovery mode is used as unconstrained, constrained or a combination of both cases. For example, in biomedical applications, SMPs will experience constraint from the surrounding tissues and will not be subjected to free recovery. The cyclic test procedure can be presented as a two-dimensional (ε–σ) or three-dimensional (ε–*T*–σ) plots (Figure 3.16). The T_{trans} can only be determined from the three-dimensional graph. The shape memory properties are influenced by various thermomechanical test parameters such as applied stress and strain, strain rate, cooling and heating rate, deformation and cooling temperature, number of cycles [51].

The thermomechanical cyclic test procedure comprises four steps. In step 1, the specimen is stretched to a strain value ε_m with an ultimate stress σ_m above T_{trans}. In step 2, it is cooled to a low temperature (below T_{trans}) at constant stress or strain to fix the deformed shape. Step 3 is a stress-free stage where the specimen experiences a spring back, ε_u is a residual strain at the fixed temporary shape. In step 4, the sample is reheated (typically close to or above T_{trans}) to recover the original shape, ε_p is the residual strain after shape recovery. From the stress–strain curve of the thermomechanical test, the shape memory parameters namely shape recovery (R_r) and shape fixity (R_f) can be calculated from the following equations[42, 52]:

$$R_r(N) = \varepsilon_m - \varepsilon_p(N) \,/\, \varepsilon_m - \varepsilon_p(N-1)$$

$$R_f(N) = \varepsilon_u(N) \,/\, \varepsilon_m$$

where *N* is the number of cycles.

3.5 Effect of thermomechanical cyclic conditions on SME

SME is not an intrinsic property of a single polymer; it results from a combination of the polymer molecular structure, morphology, processing, and programming and recovery conditions [44, 45, 51]. The shape memory behaviors of SMP depend on the different phase states and mechanical properties above and below the T_g. The shape memory properties vary by changing the conditions of thermomechanical cyclic test. The programming conditions like deformation temperature, deformation rate, maximum strain, cooling temperature and cooling time have the strong influence on the SME. The recovery parameters such as recovery temperature and recovery time have also the effect on the shape memory characteristic. In addition, cooling and heating rate also have the effect on the shape memory function of SMP. A slow cooling rate results in a higher relaxation of the stress at higher temperatures and leads to a lower σ_{max} as compared to a faster cooling rate. A slower heating rate leads to a higher σ_{max} and a shape recovery at lower temperature because it gives enough time for an effective release of the stored strain energy. In order to design effective devices made of thermoresponsive SMP, it needs to understand the effect of the conditions of each step of thermomechanical cyclic test on the SME. A precisely controlled shape fixity and faster shape recovery is essential for the potential application of SMPs.

3.5.1 Effect of programming conditions

3.5.1.1 Effect of deformation temperature

SMPs can be deformed below T_{trans} or above T_{trans}. The deformation temperature has a strong influence on the SME of SMPs. The functional characteristics of SMPs such as rigidity, elastic modulus and rubbery modulus drastically change below and above T_{trans}. When SMPs deformed below T_{trans}, the energy lost is high due to high molecular friction, as a result, affect the shape memory function. SMPs deformed below T_{trans} (cold stretching) usually exhibited lower shape fixity and faster shape recovery. Faster shape recovery can be obtained by deforming SMPs at a lower temperature and recovery at a higher temperature. The effect of deformation temperature on SME of a thermoplastic semicrystalline multiblock copolymer has been reported by Yan et al. [52]. They prepared the shape memory multiblock copolymer from crystallizable poly(ε-caprolactone) (PCL) and crystallizable poly(3*S*-isobutyl-morpholin-2,5-dione) (PIBMD). PIBMD crystals act as permanent physical netpoints as its high T_m of 170 °C, whereas both the PCL crystalline phase and the PIBMD amorphous phase act as switching domains as the T_m (38 °C) of PCL crystallites is close to the T_g (42 °C) of PIBMD. The PCL-PIBMD SMP was deformed at 20 and 50 °C, and observed that the PCL-PIBMD deformed at 50 °C exhibited the higher shape fixity and lower maximum stress than deformed at 20 °C (Figure 3.17). Both PCL and PIBMD amorphous phases

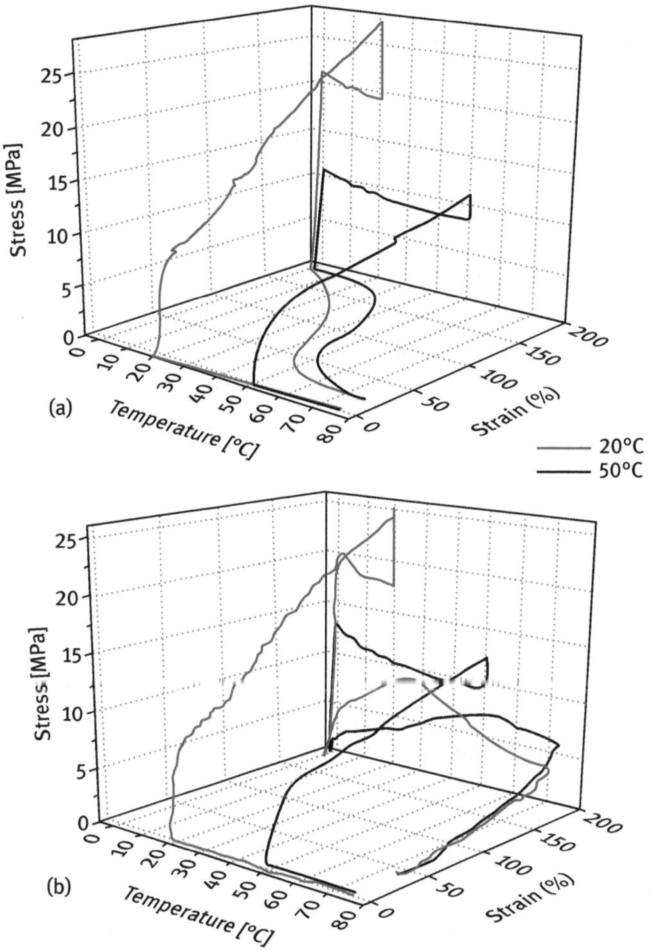

Figure 3.17: The first cycle curves of cyclic tests of PCL-PIBMD films programmed to 200% at 50 and 20 °C, respectively, with the strain rate of 1 mm/min and recovered under (a) stress-free and (b) constant-strain conditions.

can be oriented when the sample was deformed at 50 °C. Therefore, both PCL crystals and PIBMD amorphous phase can contribute to the fixation of the temporary shape because PCL crystals and PIBMD amorphous phase were fixed by crystallization and vitrification, respectively, when cooled to 0 °C. The PIBMD amorphous phase remained in the glassy state when the PCL-PIBMD samples were programmed at 20 °C; therefore, the temporary shape was fixed mainly by the crystallization of PCL domains. They demonstrate that both macromolecular structure and thermomechanical investigations are useful for understanding of complicated shape memory mechanisms in multiblock copolymer containing two crystallizable segments. Azra et al. [53] studied the influence of deformation temperature on the SME of polyurethanes. The recoverable strain observed to be decreased strongly as deformation temperatures increased above T_g +20 °C. This can be attributed to the modification or destruction of the primary network at high deformation temperature; as a consequence, the hard domains have relatively low mechanical stability, leading to a loss of memory of the primary shape. Similar observation was obtained by Hu et al. [54]. It was demonstrated that the shape memory polyurethanes exhibited SME at deformation temperature ranging from T_g –10 to T_g +50 °C, and showed better SME at deformation temperature ranging from T_g to T_g +25 °C. It was suggested that shape memory polyurethanes exhibited good SME over the deformation temperature range in which the macromolecular segments of the reversible phase have gained sufficient mobility when responding to the external stress, while the fixed phase is able to stay relatively stable.

3.5.1.2 Effect of deformation rate

Deformation rate has a significant effect on the SME of SMPs. At high deformation rate, the material may change from rubbery to leathery and eventually to glassy state. The orientation of macromolecular chains may differ as a function of deformation rate. A better shape recovery is observed when SMP is deformed at a faster deformation rate. The stress loss during relaxation is higher for a lower deformation rate. The effect of deformation rate on the SME of Veriflex-E has been studied by McClung et al. [55]. The specimen was subjected to different deformation rates of 0.5, 5, and 50 mm/min. The strain recovery was found to increase with the increase of the deformation rate. At higher deformation rate, the material experiences less time for configurational changes to take place at elevated temperature, resulting in the less resistance of the material produces against locking in the strain and exhibits the better strain recovery. Hu et al. [54] also observed the same results for the shape memory polyurethanes. The cyclic test was carried out at different deformation rates of 2, 10 and 50 mm/min. The recovery ratios were observed to be increased with increasing deformation rate. At a low deformation rate, a large amount of stress is lost through stress relaxation, resulting in the decrease of recovery ratio.

3.5.1.3 Effect of maximum strain

Maximum strain applied during programming condition has also an effect on the shape memory properties of SMPs. Generally the shape fixity increases with the increase in maximum strain, whereas the shape recovery decreases with increase in maximum strain. The effect of maximum strain on the shape memory properties of segmented polyurethanes have been demonstrated [35]. SME was tested with different maximum strains of 50%, 100%, 150%, 200% and 250%. It was observed that the shape fixity increased with the increase in maximum strain. The shape fixity was much lower at maximum strain of 50% as compared to maximum strain of 100%. This might be because the soft segments were not fully extended to 50% maximum strain, which resulted in no much strain-induced crystallization. This implied that the segmented polyurethanes had to be extended to at least over 100% maximum strain to achieve high shape fixity. However, the shape recovery found to decrease with increasing maximum strain. This is because of change in HS domains from isolated to interconnected state due to the larger maximum strain. Hu et al. [54] also studied the effect of maximum strain of 100%, 200% and 300% on the SME of segmented polyurethanes. They also observed the similar results that the shape recovery ratio decreased with the increase in maximum strain. This is due to the substantial deformation of HSs during large elongation.

3.5.1.4 Effect of cooling temperature

During programming, SMP is cooled below T_{trans} to fix the temporary shape. The cooling temperature has an effect on the stiffness, and crystallinity of the material leads to the effect on shape memory properties of SMPs. Choi et al. [56] investigated the effect of cooling temperature on the SME of acrylic-based SMP (T_g = 55 °C). Shape fixing was performed at different cooling temperatures of 10, 20, 30, 40 and 50°C. The shape fixity and shape recovery increased with the increase in cooling temperature from 10 to 50 °C. However, SMPs fixed at 10 °C showed very poor SME. This is due to the development of high fixation stress (σ_{fix} = 30 MPa) during cooling at very low temperature resulting in the partial destruction of networks. This indicates that a very low cooling temperature should be avoided to obtain good shape memory performance of this amorphous network. It is better to opt for cooling temperature just below T_g to avoid a high increase in stress during cooling and as a consequence minimizes the irreversible bond breakage.

3.5.1.5 Effect of cooling time

The crystallization or vitrification of the SMP during cooling is influenced by the cooling time. SMPs exhibit good shape fixity when the soft segments of SMPs get sufficient crystallization during cooling time. When the SMPs are fixed at lower cooling time, a larger fraction of the deformation stress is relaxed rather than stored. The effect of cooling time on the SME of segmented polyurethane ionomers has been studied by Zhu et al. [57]. SME was evaluated at the cooling time of 30, 150, 300, 900 s. The shape fixity ratio increased with the increase in cooling time from 30 to 300 s, and after that it did not show any effect on the shape fixity ratio. This indicates that the soft segments of SMPs achieve sufficient time for crystallization during the cooling time from 30 to 300 s. However, the shape recovery ratio decreased with the increase in cooling time from 300 to 900 s. This may be due to the destruction of network structure during the overcooling of SMP.

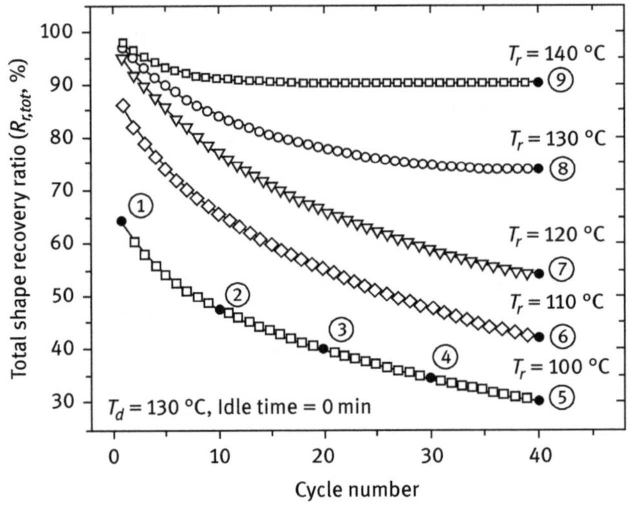

Figure 3.18: Shape recovery ratio plot as a function of recovery temperature and cycle time. Hold time is 300 s, T_d = 130 °C and idle time between two SM cycles is zero.

3.5.2 Effect of recovery conditions

3.5.2.1 Effect of recovery temperature

The frozen macromolecular chains become reactivated when the temporary shape is reheated and recover the original shape. The recovery temperature has an effect on the SME of SMP. SMPs show better shape recovery behavior when the entire frozen macromolecular chains are reactivated and the strain energy dissipations are fast. The effect of recovery temperature on the shape recovery of epoxy-based thermosetting SMP (T_g = 95 °C) has been investigated [58]. The recovery ratio was evaluated at different recovery temperatures of 100, 110, 120, 130 and 140 °C. The recovery ratio was observed to be increased with the increase of recovery temperature (Figure 3.18). The accumulated residual strain energy within the macromolecular networks lead to the gradual decrease in recovery ratio at a lower recovery temperature. Higher recovery temperature assists to suppress the accumulation of residual strain energy in the polymer network and as a result enhances the recovery ratio. Hu et al. [54] studied the effect of recovery temperature on SME of segmented polyurethanes. The recovery behavior was performed at the recovery temperature of T_g, T_g + 5, T_g +15 and T_g +25 °C. Similarly the recovery ratio increased with the increase of recovery temperature as observed by Yu et al. A low temperature cannot reactivate the large frozen molecular segments that still retain some deformation. When the recovery temperature is increased, then more and more large frozen molecular segments are reactivated and the recovery ratio is increased. Lan et al. [59] also investigated the effect of recovery temperature on the SME of styrene-based thermoset SMP composites (T_g = 64 °C). The recovery effect was investigated at the recovery temperature of T_g –20 °C, T_g –10 °C, T_g

and T_g +10 °C. The material exhibited faster recovery at the recovery temperature of T_g and T_g +10 °C. This is due to the fast strain energy dissipation in the SMP. The shape recovery was found to be very slow at the recovery temperature of below T_g. This is because of the partial strain energy dissipation in the SMP. However, the material did not show SME below the recovery temperature of T_g –20 °C. Isothermal shape recovery behavior of SMP has also been investigated [44]. It is important to study for the biomedical applications of SMP, as most of the thermoresponsive SMP devices designed for recovery at the body temperature and the recovery function are inherently isothermal.

3.5.2.2 Effect of recovery time

The recovery time is also a recovery parameter that influences the recovery behavior of SMP. The SMP should be reheated to enough time for effective releasing of the stored energy. The recovery will occur at higher recovery temperature, when the recovery time is less. Hu et al. [54] studied the effect of recovery time (10 and 40 min) on shape recovery of polyurethanes. It was found that the recovery was increased with the increase of recovery time. It was suggested that a sufficient recovery time is needed for the frozen molecular segments to reactivate.

References

[1] Karak N. Fundamentals of polymers. New Delhi: PHI learning private limited, 2009.

[2] Cheng HN, Early TA. NMR studies of polymeric materials-an overview. Macromol Symp 1994;86:1–14.

[3] Bertmer M, Buda A, Blomenkamp-Höfges I, Kelch S, Lendlein A. Biodegradable shape-memory polymer networks: characterization with solid-state NMR. Macromolecules 2005;38(9):3793–9.

[4] Powers DS, Vaia RA, Koerner H, Serres J, Mirau PA. NMR characterization of low hard segment thermoplastic polyurethane/carbon nanofiber composites. Macromolecules 2008;41(12):4290–5.

[5] Eisenbach CD, Gronski W. Hydrogen bonding and phase separation in segmented polyurethane elastomers as studied by ^{13}C NMR magic angle spinning and FT-IR spectroscopy. Macromol Rapid Commun 1983;4:707–13.

[6] Kalita H, Karak N. Mesua ferrea L. seed oil-based hyperbranched shape memory polyurethanes: effect of multifunctional component. Polym Eng Sci 2012;52:2454–61.

[7] Snively CM, Koenig JL. Polymer applications of IR and Raman spectroscopy. Encycl Spectrosc Spectrom 2010;2213–20.

[8] Mattia J, Painter P. A comparison of hydrogen bonding and order in a polyurethane and poly(urethane-urea) and their blends with poly(ethylene glycol). Macromolecules 2007;40(5):1546–54.

[9] Wang CB, Cooper SL. Morphology and properties of segmented polyether polyurethaneureas. Macromolecules 1983;16(5):775–86.

[10] Dai Z, Yang K, Dong Q. Mechanical, thermal and morphology properties of thermoplastic polyurethane copolymers incorporating of varying poly(propyleneoxide) molecular weight. Open J Synth Theory Appl 2015;4:41–57.
[11] Yilgor I, Yilgor E, Guler IG, Ward TC, Wilkes GL. FTIR investigation of the influence of diisocyanate symmetry on the morphology development in model segmented polyurethanes. Polymer 2006;47(11):4105–14.
[12] Kasai N, Kakudo M. X-ray diffraction by macromolecules. Tokyo: Kodansha, 2005.
[13] Seidel A. Characterization and analysis of polymers. New Jersey: John Wiley & Sons, 2008.
[14] Chu B, Hsiao BS. Small-angle X-ray scattering of polymers. Chem Rev 2001;101(6):1727–61.
[15] Koberstein JT, Stein RS. Small-angle X-ray scattering studies of microdomain structure in segmented polyurethane elastomers. J Polym Sci Polym Phys Ed 1983;21(8):1439–72.
[16] Das S, Yilgor I, Yilgor E, Inci B, Tezgel O, Beyer FL, et al. Structure e property relationships and melt rheology of segmented, non-chain extended polyureas: effect of soft segment molecular weight. Polymer 2007;48:290–301.
[17] Chu B, Gao T, Li Y, Wang J, Desper CR, Byrne CA, et al. Microphase separation kinetics in segmented polyurethanes: effects of soft segment length and structure. Macromolecules 1992;25(21):5724–9.
[18] Goldstein J, Newbury DE, Joy DC, Lyman CE, Echlin P, Lifshin E, et al. Scanning electron microscopy and X-Ray microanalysis. New York: Springer, 1981.
[19] Jinnai H, Spontak RJ, Nishi T. Transmission electron microtomography and polymer nanostructures. Macromolecules 2010;43(4):1675–88.
[20] Libera MR, Egerton RF. Advances in the transmission electron microscopy of polymers. Polym Rev 2010;50(3):321–39.
[21] Serrano M, MacKnight WJ, Thomas EL, Ottino JM. Transport-morphology relationships in segmented polybutadiene polyurethanes: 1. Experimental results. Polymer 1987;28(10):1667–73.
[22] Koutsky JA, Hien N V., Cooper SL. Some results on electron microscope investigations of polyether-urethane and polyester-urethane block copolymers. J Polym Sci Part B Polym Lett 1970;8(5):353–9.
[23] Li C, Cooper SL. Direct observation of the micromorphology of polyether polyurethanes using high-voltage electron microscopy. Polymer 1990;31(1):3–7.
[24] Magonov SN, Reneker DH. Characterization of polymer surfaces with atomic force microscopy. Annu Rev Mater Sci 1997;27(1):175–222.
[25] Maver U, Maver T, Persin Z, Mozetic M, Vesel A, Gaberscek M, et al. Polymer characterization with the atomic force microscope. London: INTECH Open Access Publisher, 2013:113–32.
[26] Cao Q, Liu P. Structure and mechanical properties of shape memory polyurethane based on hyperbranched polyesters. Polym Bull 2006;57(6):889–99.
[27] Aneja A, Wilkes GL. A systematic series of "model" PTMO based segmented polyurethanes reinvestigated using atomic force microscopy. Polymer 2003;44(23):7221–8.
[28] Schön P, Bagdi K, Molnár K, Markus P, Pukánszky B, Julius Vancso G. Quantitative mapping of elastic moduli at the nanoscale in phase separated polyurethanes by AFM. Eur Polym J 2011;47(4):692–8.
[29] Koenig JL. Infrared and Raman spectroscopy of polymers. Ohio: Smithers Rapra, 2001.
[30] Edwards HGM, Johnson AF, Lewis IR. Applications of Raman spectroscopy to the study of polymers and polymerization processes. J Raman Spectrosc 1993;24(8):475–83.
[31] Leim H, Yeung LY. Segment self-orientational behavior in shape memory polymer thin films probed by Raman spectroscopy. J Appl Polym Sci 2007;105(2): 765–70.
[32] Parnell S, Min K, Cakmak M. Kinetic studies of polyurethane polymerization with Raman spectroscopy. Polymer 2003;44(18):5137–44.

[33] Müller AJ, Michell RM. Differential scanning calorimetry of polymers. New Jersey: Wiley, 2016:72–99.

[34] Kong Y, Hay JN. The measurement of the crystallinity of polymers by DSC. Polymer 2002;43:3873–8.

[35] Ji FL, Hu JL, Li TC, Wong YW. Morphology and shape memory effect of segmented polyurethanes. Part I: with crystalline reversible phase. Polymer 2007;48(17):5133–45.

[36] Menczel JD, Prime RB. Thermal analysis of polymers: fundamentals and applications. New Jersey: Wiley, 2009.

[37] Menard KP. Dynamic mechanical analysis: a practical introduction. CRC press, 1999.

[38] Merline JD, Nair CPR, Gouri C, Bandyopadhyay GG, Ninan KK. Polyether polyurethanes: synthesis, characterization, and thermoresponsive shape memory properties. J Appl Polym Sci 2008;107(6):449–56.

[39] Kim BK, Shin YJ, Cho SM, Jeong HM. Shape-memory behavior of segmented polyurethanes with an amorphous reversible phase: the effect of block length and content. J Polym Sci Part B Polym Phys 2000;38(20):2652–7.

[40] Jeong HM, Kim B, Lee SY. Miscibility and shape memory property of poly(vinyl chloride)/thermoplastic polyurethane blends. J Mater Sci 2011;36(22):5457–63.

[41] Merline JD, Nair CPR, Gouri C, Shrisudha T, Ninan KN. Shape memory characterization of polytetra methylene oxide/poly (acrylic acid-co-acrylonitrile) complexed gel. J Mater Sci 2007;42(15):5897–902.

[42] Heuchel M, Sauter T, Kratz K, Lendlein A. Thermally induced shape-memory effects in polymers: quantification and related modeling approaches. J Polym Sci Part B Polym Phys 2013;51(8):621–37.

[43] Liu C, Qin H, Mather PT. Review of progress in shape memory polymers. J Mater Chem 2007;17(16):1543.

[44] Messori M, Degli Esposti M, Paderni K, Pandini S, Passera S, Ricco T, et al. Chemical and thermomechanical tailoring of the shape memory effect in poly(ε-caprolactone)-based systems. J Mater Sci 2013;48(1):424–40.

[45] Yakacki CM, Shandas R, Safranski D, Ortega AM, Sassaman K, Gall K. Strong, tailored, biocompatible shape-memory polymer networks. Adv Funct Mater 2008;18(16):2428–35.

[46] Kalita H, Karak N. Bio-based elastomeric hyperbranched polyurethanes for shape memory application. Iran Polym J 2012;21(4):263–71.

[47] Kalita H, Karak N. Biobased hyperbranched shape-memory polyurethanes: effect of different vegetable oils. J Appl Polym Sci 2014;131(1):1–8.

[48] Li F, Larock RC. New soybean oil-styrene-divinylbenzene thermosetting copolymers. V. Shape memory effect. J Appl Polym Sci 2002;84(8):1533–43.

[49] Kong X, Liu G, Qi H, Curtis JM. Preparation and characterization of high-solid polyurethane coating systems based on vegetable oil derived polyols. Prog Org Coatings 2013;76(9):1151–60.

[50] Sauter T, Heuchel M, Kratz K, Lendlein A. Quantifying the shape-memory effect of polymers by cyclic thermomechanical tests. Polym Rev 2013;53(1):6–40.

[51] Santiago D, Ferrando F, De La Flor S. Effect of different shape-memory processing methods on the thermomechanical cyclic properties of a shape-memory polyurethane. J Mater Eng Perform 2014;23(7):2561–6.

[52] Yan W, Fang L, Noechel U, Kratz K, Lendlein A. Influence of deformation temperature on structural variation and shape-memory effect of a thermoplastic semi-crystalline multiblock copolymer. Express Polym Lett 2015;9(7):624–35.

[53] Azra C, Plummer CJG, Månson J-AE. Isothermal recovery rates in shape memory polyurethanes. Smart Mater Struct 2011;20(8):82002.

[54] Hu JL, Ji FL, Wong YW. Dependency of the shape memory properties of a polyurethane upon thermomechanical cyclic conditions. Polym Int 2005;54(3):600–5.
[55] McClung AJ, Tandon GP, Baur JW. Deformation rate, hold time, and cycle-dependent shape memory performance of Veriflex-E resin. Mech Time-Depend Mater 2013;17(1):39–52.
[56] Choi N, Lendlein A. Degradable shape-memory polymer networks from oligo[(l-lactide)-ran-glycolide]dimethacrylates. Soft Matter 2007;3(7):901–9.
[57] Zhu Y, Hu J, Yeung K. Effect of soft segment crystallization and hard segment physical crosslink on shape memory function in antibacterial segmented polyurethane ionomers. Acta Biomater 2009;5(9):3346–57.
[58] Yu K, Li H, McClung AJW, Tandon GP, Baur JW, Qi HJ. Cyclic behaviors of amorphous shape memory polymers. Soft Matter 2016;12(13):3234–45.
[59] Lan X, Liu Y, Lv H, Wang X, Leng J, Du S. Fiber reinforced shape-memory polymer composite and its application in a deployable hinge. Smart Mater Struct 2009;18(2):24002.

4 Shape memory polyurethanes: From materials to synthesis

4.1 Introduction

Polyurethanes are linear polymers that have a molecular backbone containing carbamate groups (–NHCOO–). These groups, called urethane, are produced through a rearrangement reaction between a diisocyanate and a polyol. Otto Bayer developed urethane chemistry in 1937 at the I.G. Farben Laboratories [1, 2]. Polyols used in the preparation of polyurethanes are of two types, viz. macroglycol and chain extender. The properties of polyurethanes depend on the molecular weight, structure, reactivity of the reactants and NCO/OH ratio of the composition. Among different shape memory polymers, shape memory polyurethanes are the most versatile due to their biocompatibility, wide-range tunable stiffness, large deformation, large recovery, good elastic property, water vapor permeability and multiresponsive shape memory effect (SME) [3, 4]. The shape memory polyurethanes are made up of alternating hard and soft segments. The hard segment is composed of diisocyanate with low-molecular-weight diol or diamine chain extender; hence, it is rigid. The soft segment of the polyurethanes is usually moderately high-molecular-weight long-chain polyol that is flexible in nature. Owing to the polar character of urethane groups, the hard urethane domains are more hydrophilic in contrast to the soft polyol domains that cause the microphase separation in the structure. The flexible soft segments are responsible for the reversible phase transformation, i.e., switching segment that allows for the SME, while the hard segments are responsible for the memorizing of permanent shape [4–7]. The SME as well as the performance of shape memory polyurethanes can be tailored by judicious variation of chemical constituents, molar ratio of hard to soft segments and polymerization process [3, 8, 9]. In addition, chain length, molecular weight, conformation, arrangement, crystallization of soft segment and morphological structure play an important role to the tune of shape memory properties of polyurethanes. Further, the structure of hard segments, i.e., structure of diisocyanate and chain extender has also the strong influence on the shape memory properties of polyurethanes. The brief descriptions of these components are presented below.

4.2 Materials

4.2.1 Diisocyanate

Isocyanates are highly reactive compounds and can readily react with the groups containing active hydrogen. The high reactivity of the isocyanate (–N=C=O) is due to its two cumulated double bond. Isocyanate is a first essential component in the for-

https://doi.org/10.1515/9783110570175-004

mation of polyurethanes that act as hard segment. The properties of shape memory polyurethanes largely depend on the structure, shape, reactivity and crystallization of the isocyanate. The reactivity of the diisocyanates depends on the position of the isocyanate group, substituents, symmetry, steric effect and ring structure. In toluene diisocyanate (TDI), the isocyanate group at para position is more reactive than the isocyanate group at ortho position [10]. The bulky or branched substituent near the –NCO group retards the reactivity. In 2,6-toluene diisocyanate, once the one –NCO reacted and then the reactivity of other one decreases due to the steric effect. The presence of electron-withdrawing substituent increases the reactivity by increasing the partial positive charge on the isocyanate carbon atom. The symmetrical diisocyanate has the tendency to form crystallizable structure to some extent that decreases the reactivity. Utilizing the different diisocyanates, the properties of the polyurethanes can be tuned. Diisocyanate can be either aromatic or aliphatic. The aromatic diisocyanates are more reactive than aliphatic ones. The most commonly used diisocyanates for shape memory polyurethane synthesis are listed in Table 4.1. The most widely used diisocyanates for the preparation of shape

Table 4.1: Diisocyanates used for the synthesis of shape memory polyurethanes.

Name of diisocyanate	**Chemical structure**
2,4- and 2,6-toluene diisocyanate	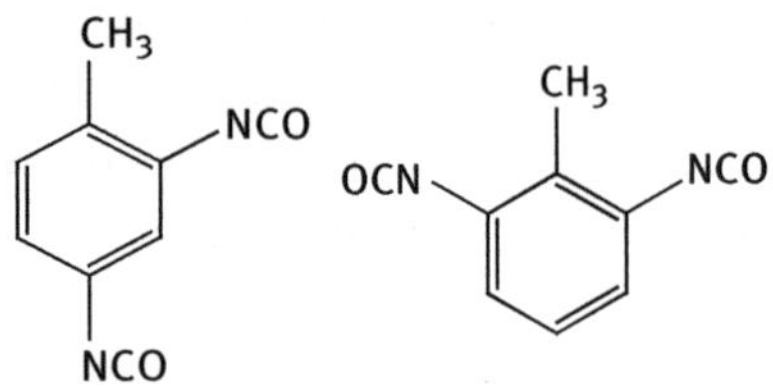
4,4′-Methylenediphenyl diisocyanate	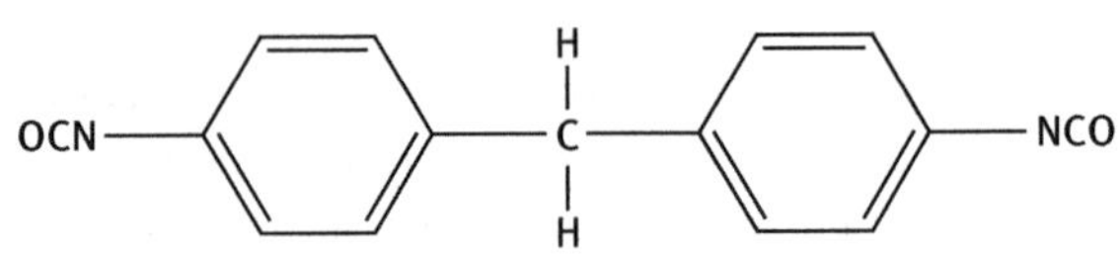
3,3′-Dimethyl-diphenylmethane-4,4′diisocyanate	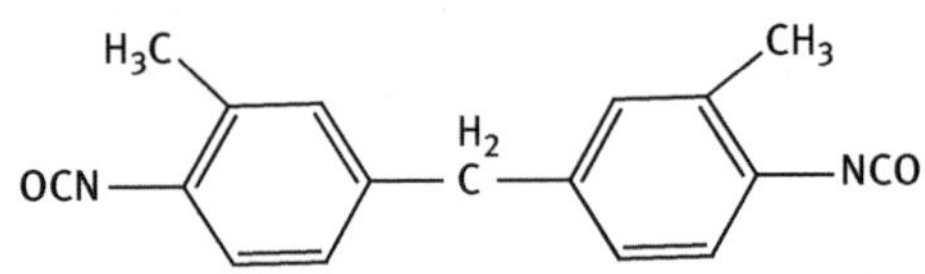
1,5-Naphthalene diisocyanate	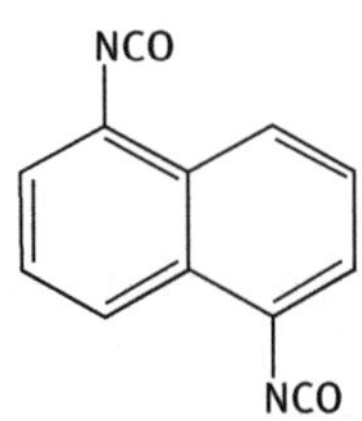

Table 4.1: (continued)

Name of diisocyanate	**Chemical structure**
1,6-Hexamethylene diisocyanate	$OCN—(CH_2)_6—NCO$
1,4-Cyclohexyl diisocyanate	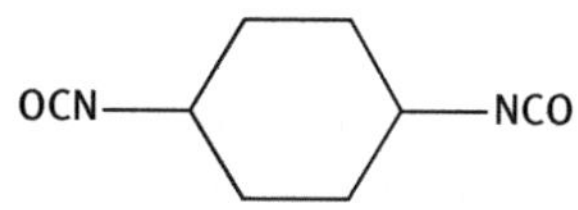
4,4′-Dicyclohexylmethane diisocyanate	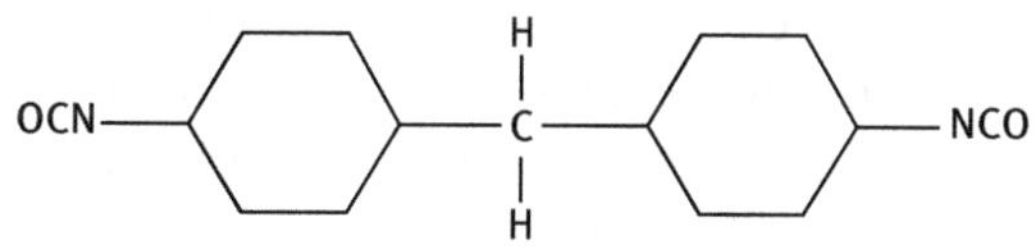
Isophorone diisocyanate	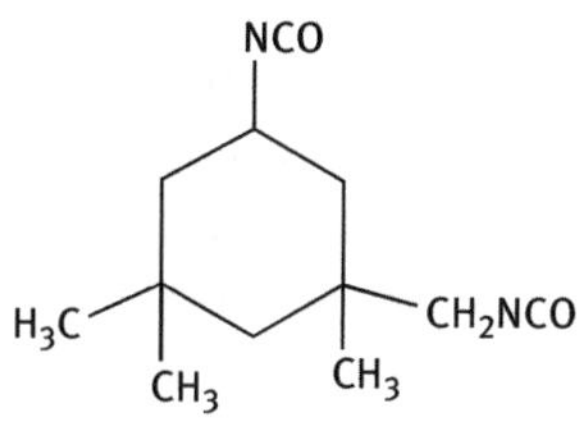
Norbornene diisocyanate	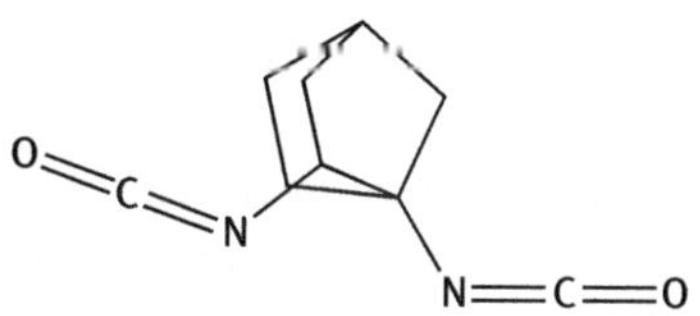
p-Phenylene diisocyanate	

memory polyurethanes are TDI and 4,4′-diphenylmethane diisocyanate (MDI). In most of the cases, TDI used is a mixture of 2,4- and 2,6-isomers in 80:20 molar ratio. Similarly, MDI has three isomers, namely 4,4-, 2,4- and 2,2-diphenylmethane diisocyanates. However, 4,4-isomer is used in most of the commercial polyurethanes. Though aromatic diisocyanates are more reactive than aliphatic one, polyurethanes obtained from the aromatic diisocyanates have lower oxidation and weaker ultraviolet stabilization than the polymers with aliphatic diisocyanates [2]. However, the aromatic diisocyanate-based shape memory polyurethanes exhibited good mechanical strength and shape memory behaviors. The rigidity of the polyurethane network is supposed to increase with the following order: aliphatic < cycloaliphatic < aromatic [11]. More symmetrical structure of diisocyanate and a chain extender increases the formation of organized structures and crystallinity, leading to more complete phase segregation which in turn offer the better modulus, stiffness and shape memory properties.

Desai et al. [12] reported that TDI-based polyurethanes exhibited higher tensile strength and lower elongation as compared to MDI-based polyurethanes irrespective of the type of polyol, even though both have aromatic structure. MDI has a linear symmetric structure consisting of two aromatic rings, giving lower tensile strength owing to the lack of rigidity in the backbone. However, in the case of TDI a higher rigidity is provided due to the direct association of the two isocyanate groups with the same phenyl ring. The effect of structure of diisocyanates on the properties of polyurethanes was studied by Javni et al. [11]. They reported that aromatic triisocyanates imparted the highest density, glass transition, modulus and tensile strength, but exhibited the lowest elongation at break, swelling and impact resistance. Aliphatic triisocyanates and diisocyanates conferred rubbery materials with the highest elongation at break, highest swelling and the lowest tensile strength. However, polyurethanes with aromatic and cycloaliphatic diisocyanates were similar in properties, with values present between the aromatic triisocyanates and aliphatic triisocyanates. Pandya et al. [13] also studied the effect of different diisocyanates (TDI, MDI crude, HDI [hexyl diisocyanate] and IPDI [isophorone diisocyanate]) on the properties of polyurethanes. They also found that TDI-based polyurethane showed higher tensile strength among the others. This is due to the direct association of isocyanate group with phenyl ring, which imparted the higher rigidity of the structure. In addition, the higher reactivity of TDI which results from delocalization of negative charge on –NCO by aromatic structure is also responsible for higher tensile strength. The IPDI-based polyurethane exhibited lower mechanical properties as compared to the TDI-based one. This is because of lack of delocalization of the negative charge on –NCO, thus reducing the reactivity of IPDI. The rigidity in IPDI is moderate due to the nonplanar structure (chair configuration) as all cyclic carbon atoms are sp^3 hybridized. The substituent on the cyclohexane ring reduces the symmetry as a result of a general decrease in tensile and hardness properties. The HDI-based polyurethane showed higher modulus when compared to the others. This is because of the presence of hydrogen bonding between two polymeric chains, which is facilitated by the even number of $-CH_2$ groups present in diisocyanate as well as in the chain extender. The $-CH_2$sequence of HDI together with that of butanediol forms a tight crystalline structure, thus it exhibits higher moduli and hardness.

The effect of planner and bend shape diisocyanates on the SME of polyurethanes has been reported by Yang et al. [14]. They have used 1,6-diphenyldiisocyanate (PDI) as planar diisocyanate, while 4,4′-diphenylmethyldiisocyante as bend diisocyanate. It was seen that PDI-based polyurethane showed better shape recovery than MDI-based one. This is due to the tighter interactions among the PDI-based hard segment because of its planar structure. Ping et al. [15] investigated the effect of hard segments on shape memory behaviors of segmented polyurethanes prepared from HDI, MDI, TDI and IPDI. It was observed that HDI- and MDI-based polyurethanes exhibited higher shape recovery forces as compared to the TDI and IPDI one. This can be attributed to the higher degree of microphase separation as a result of the crystallization of HDI- and MDI-based hard segments. The effect of hard segment arrangement on the SME of segmented polyurethanes was studied by Ji et al. [16]. They prepared shape memory polyurethanes having interconnected, isolated and no hard-segmented domains using MDI, 1,4-butanediol (BD) and PCL. The shape memory polyurethane having isolated hard segments showed better shape recovery. The shape recovery of shape memory polyurethanes reduced significantly when the hard segment domains changed from isolated to interconnected state. It was demonstrated that the shape fixity decreased with the increase of hard segment content (>30 wt%). This was due to the decrease of crystallization capability of the soft segments with the increase of hard segment content. Further, it was also demonstrated that the shape memory polyurethanes should be extended to at least over 100% strain to achieve better shape fixity. The shape memory polyurethanes having hard segments content between 25 and 30 wt% showed the best SME because of enough interactions among the hard segments to store more elastic energy. The shape memory polyurethane with low hard segment content (<20 wt%) showed poor SME because of the less interactions or physical cross-link. The similar observation was also reported by Lee et al. [8]. The shape memory polyurethanes having more than 50 wt% of hard segment did not show SME because of the excess interactions among the hard segments and the resulting rigid structure.

4.2.2 Macroglycol

Diols/polyols with average molecular weight of 500–5,000 g/mol are used as macroglycol in the preparation of shape memory polyurethane, which acts as a soft segment. Polyols are generally categorized into two groups, viz. polyether polyol and polyester polyol [10]. The structure, molecular weight and hydroxyl number of the macroglycol is an important factor in governing the ultimate properties of the shape memory polyurethanes. The architecture, chain length and content of soft segment have the effect on the SME of shape memory polyurethanes. Further, soft segments crystallization and arrangement have also influenced the SME of shape memory polyurethanes. Polyester polyols are widely used macroglycols as they exhibit excellent mechanical properties, thermal stability, outstanding tear strength and shape memory behaviors. This is due to the stronger hydrogen bond between the ester and urethane linkage as compared to the ether and urethane linkage [10, 17]. High polarity of the ester carbonyl group leads to stronger hydrogen bonding between ester and urethane linkage. Polyether-based polyurethanes show good hydrolysis resistance and higher resilience, whereas polyester-based polyurethanes show better modulus, stiffness and biodegradability. Hydroxy-terminated polybutadiene has also been used instead of polyester/polyether, where polyurethanes show excellent resistance to acid/base hydrolysis, low glass transition temperature and ability to retain elastomeric behavior. Among

Table 4.2: Macroglycols used for the synthesis of shape memory polyurethanes.

Name of macroglycol	Chemical structure
Polyethylene oxide	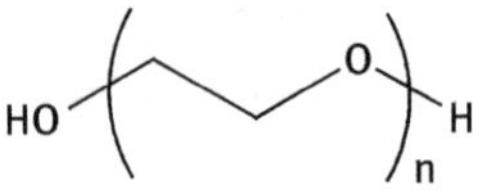
Polypropylene oxide	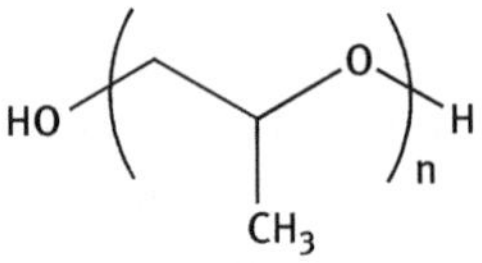
Polytetramethylene oxide	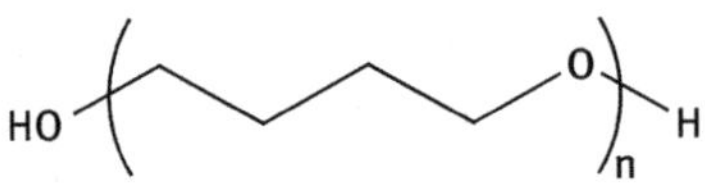
Polycaprolactone	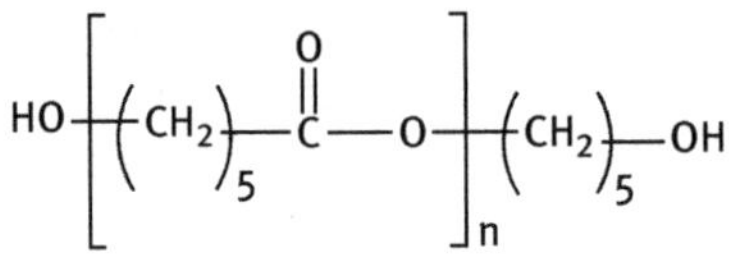
Polyethylene adipate	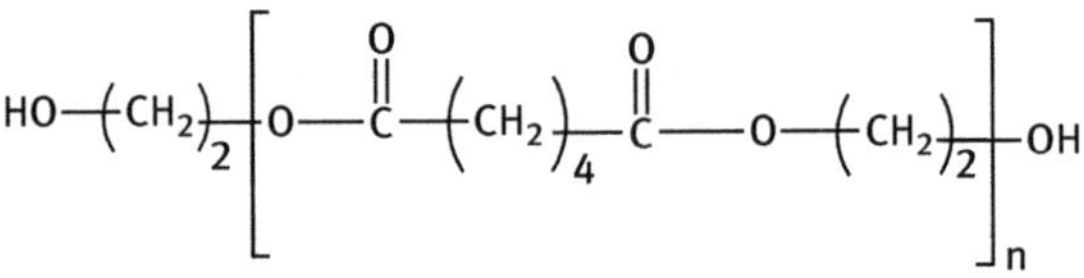
1,4-Polybutadiene diol	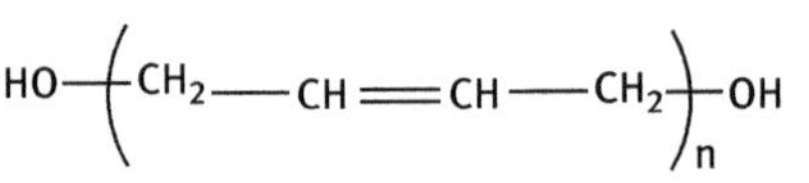

them, further the crystalline polycaprolactone (PCL) diol-based shape memory polyurethane is the most convenient to use for the shape memory applications. T_m can be used as T_{trans}. A few macroglycols used for the synthesis of shape memory polyurethanes are listed in Table 4.2.

The effect of macroglycol, viz. polypropylene glycol and hydroxyl-terminated polybutadiene on the properties of polyurethanes has been demonstrated [12]. Hydroxy-terminated polybutadiene-based polyurethane showed higher tensile strength and lower elongation as compared to polypropylene glycol-based polyurethane, because of additional cross-linking from trifunctional species. Moreover, the pendant methyl groups present in polypropylene glycol hinder the close packing of mutually attracting functional groups such as urethane/ester group and also act as internal plasticizer as a result of reducing the intermolecular forces and causing poor mechanical properties. The effect of polyol chain length (poly(tetramethyleneglycol) [PTMG; MW of 1,000 or 2,000]) on the properties of shape memory polyurethanes have been investigated by Chun et al. [9]. A long soft segment, PTMG-2000, exhibited superiority in all mechanical properties such as strain, stress and modulus because a long chain length could provide more motional freedom than a short one (PTMG-1000), forming strong interchain attractions among hard segments. Martin et al. [18] investigated the effect of macroglycol soft segment CH_2/O ratio on morphology and properties of polyurethane elastomers. They used several macroglycols such as poly(ethylene oxide) , poly(tetramethylene oxide)), poly(hexamethylene oxide), poly(octamethylene oxide), poly(decamethylene oxide) and poly(1,6-hexyl carbonate) diol for the synthesis of series of polyurethanes. It was observed that the polyurethanes prepared from macrodiols with the highest CH_2/O ratio exhibited higher hard domain crystallinity, a higher degree of phase separation, the greatest hardness and stiffness. This is due to the decrease in compatibility between the hard and soft segments with the increase in soft segment CH_2/O ratio.

The effect of soft segment molecular weight (*Mn*, PCL = 2,000, 4,000, 8,000) and soft segment content (50–90 wt%) on the SME of shape memory polyurethanes has been studied by Kim et al. [19]. Both the glass modulus and rubbery modulus found to increase with the increase in molecular weight of the soft segment. This is due to the higher soft segment and hard segments phase separation and higher soft segments crystallization. Due to the greater phase-separated structure, the hard segments stimulate physical cross-links and effectively reinforce the soft segment at the rubbery state. The higher glass state modulus reflects the higher shape fixity behavior of the shape memory polyurethane. It was observed that higher soft segment content (80 wt%) rendered a higher glassy state modulus and lower rubbery state modulus, resulting in a larger elastic modulus ratio. This replies the higher shape recovery behavior of the shape memory polyurethane. The recovery strain found to increase with the increase in molecular weight and content of soft segment. Li et al. [20] also reported the effect of soft segment molecular weight (*Mn*, PCL= 1,600–7,000) on the shape memory properties of polyurethanes. They also observed similar behaviors as reported by Kim et al. The shape recovery rate found to increase with the increase in the molecular weight of the soft segment. Ahmad et al. [21] reported the effect of different soft segments (PEG-6000 and PCL-2000) on the shape memory properties of shape memory polyurethanes. The PEG-based polyurethane showed higher shape fixity than PCL-based polyurethane because of the higher crystallinity of the PEG soft segment. PEG-based polyurethane also showed the better shape recovery ratio and faster recovery as compared to the PCL-based polyurethane. This can be attributed to the longer molecular chains of the high-molecular-weight polyol soft segment, which leads to easy deformation of the shape memory polyurethane at above T_{trans} when compared to low-molecular-weight polyol soft segment.

The effect of soft segment arrangement on the SME of shape memory polyurethanes has been reported by Cho et al. [22]. They prepared random, block and blend polyurethanes using two types of PTMG as soft segments having different molecular weight (*Mn*, PTMG = 1,000 and 1,800 g/mol). The random and blend-type shape memory polyurethanes exhibited higher shape recovery ratio as compared to the block one. However, all the copolymers showed all most same shape retention behavior. They suggested that high performance shape memory polyurethanes could be made through the precise control of the soft segment arrangement. The effect of soft segment crystallization on the shape memory function of segmented polyurethane ionomers has been studied by Zhu et al. [23]. It was observed that the shape fixity ratio increased with the increase in cooling time of the temporary shape. This implies that the crystallization time of soft segments has a significant influence on the shape fixity of shape memory polyurethanes. However, the shape recovery ratio is found to decrease slightly with the extension of cooling time.

4.2.3 Chain extender

Chain extenders are low-molecular-weight (generally, below 400 g/mol) hydroxyl or amine compounds. The choices of the chain extender and the diisocyanate determine the characteristics of hard segment and physical properties of the shape memory polyurethane [10]. Addition of chain extender to the hard segment length increases the hard segment segregation, which results in good physical properties such as enhanced modulus, stiffness and shape memory properties. Amine chain extenders are much more reactive than hydroxyl chain extenders. Diamine chain extender-based shape memory polyurethanes exhibited higher mechanical and SME than diol-based ones. This is due to the higher degree of hydrogen bonding because of the presence of urea linkage. Moreover, aromatic diamine chain extender shows better mechanical properties and higher T_g than the aliphatic one. The properties of the shape memory polyurethanes also depend on the length and architecture of the chain extender. Polymers with even diol structure adopt the lowest energy fully extended conformations that allow for hydrogen bonding in both directions perpendicular to the chain axis. However, such a hydrogen-bonding network would not be feasible for the polymers with odd diol structure in the extended conformation, and these adopt contracted, higher energy conformations. The even diol polymers have the higher possibility of crystallinity, which is the driving force for the phase separation and hence to the better properties of the polymers. The most commonly used chain extenders are mentioned in Table 4.3.

The effect of chain extenders such as bisphenol-A, bisphenol-S, bisphenol-AF and their brominated derivatives on the various properties of polyurethanes was studied by Liaw [24]. It was observed that bisphenol-S- and bisphenol-AF-based polyurethanes had higher tensile strength than bisphenol-A-based polyurethane. The brominated chain extender-based polyurethane exhibited higher flame retardancy behavior and lower thermal stability. It was found that inclusion of bromine in chain extender increases the water absorption and solvation because of its higher polarity and larger free volume. The bisphenol-S-based polyurethanes having dipolar

Table 4.3: Chain extenders used for the synthesis of shape memory polyurethanes.

Name of chain extender	Chemical structure
Ethylene glycol	
1,4-Butanediol	
1,6-Hexanediol	
Ethylene diamine	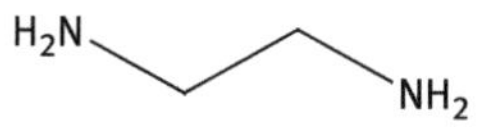
Ethanolamine	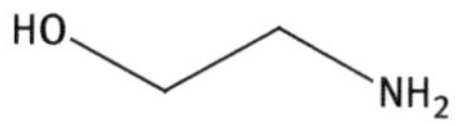
Glycerol	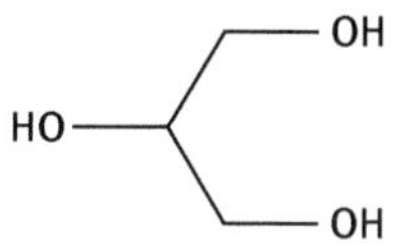
Trimethylol propane	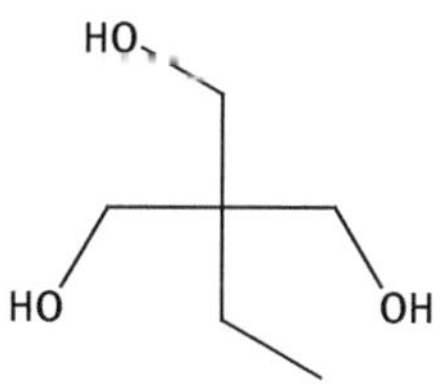
Triethanolamine	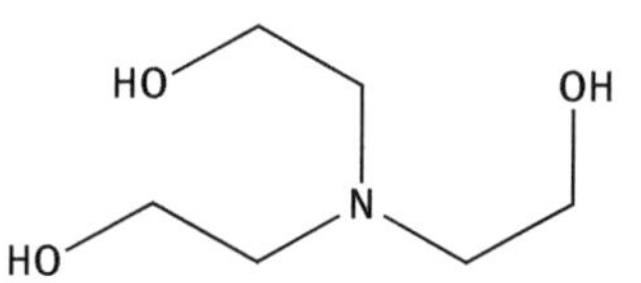
2,2-Diethyl-1,3-propanediol	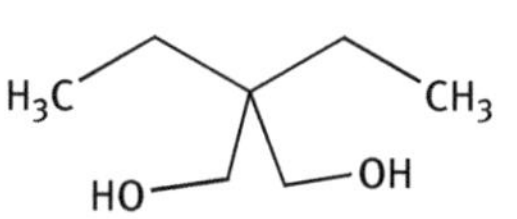
2-Butyl 2-ethyl-1,3-propanediol	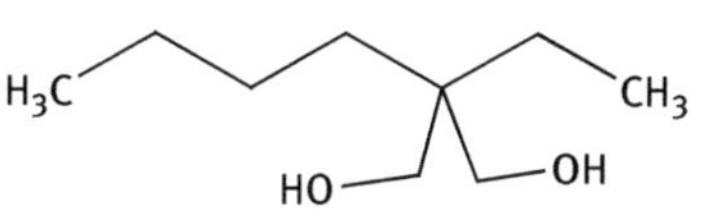
2,5-Dimethyl-3-hexyne-2,5-diol	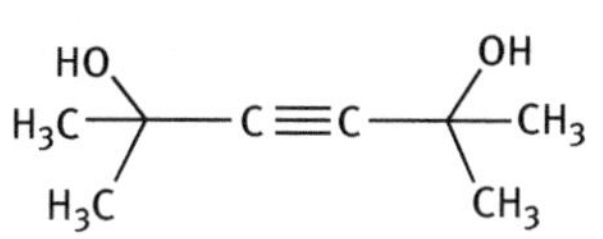

Table 4.3: (continued)

Bis[4-(β-hydroxyethoxy)] bisphenol A (Dianole-22)	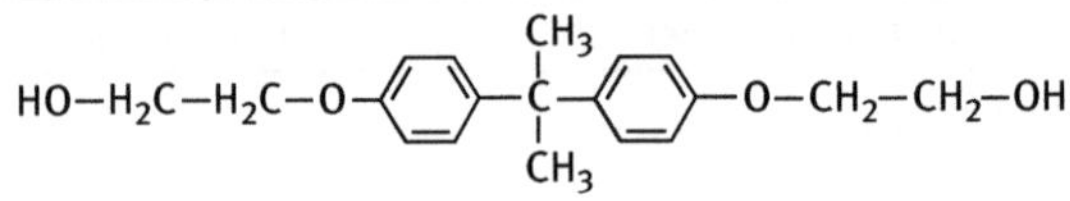
2,6-Pyridinedimethanol	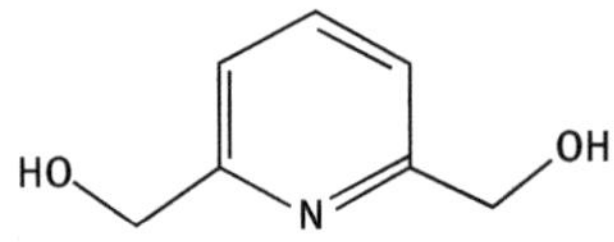
4,4′-Diaminobiphenyl	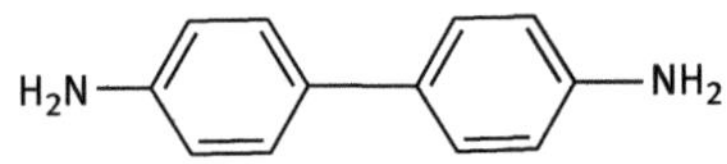
Bisphenol-S	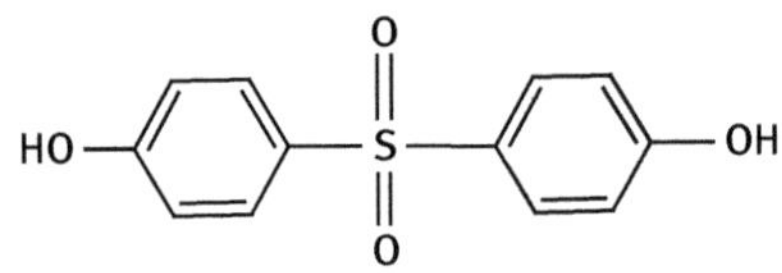
Bisphenol-A	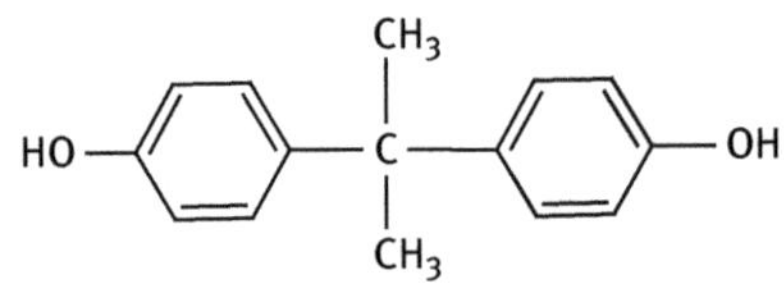
4,4′-Diaminodiphenyl sulfone	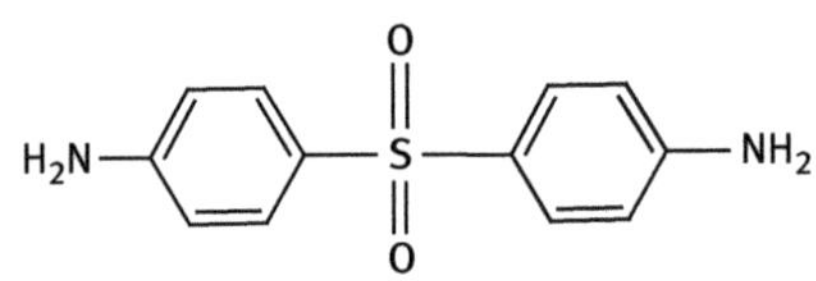
4,4′-Diaminodiphenyl methane	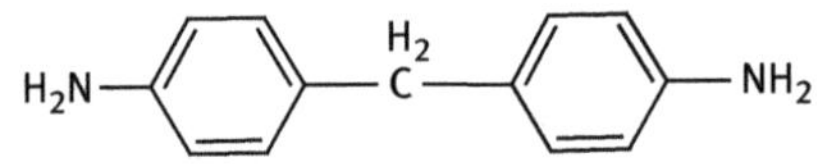
Isophthalic acid dihydrazide	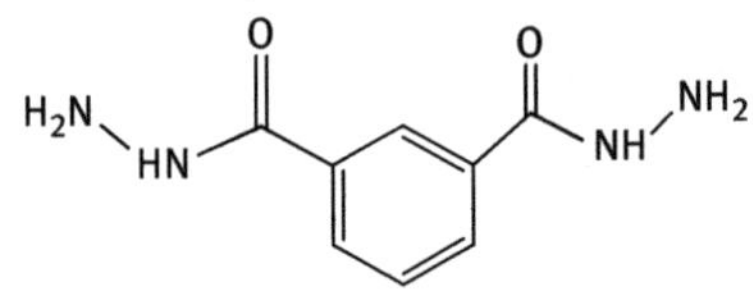
Piperazine	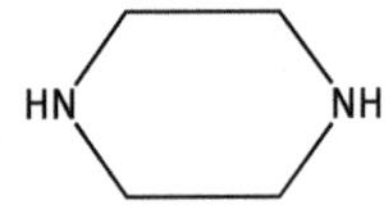

sulfonyl groups in the polymer chains display higher solvation and water absorption. The bisphenol-S-based polyester polyurethanes exhibited higher dynamic properties than the bisphenol-A- and bisphenol-AF-based polyurethanes because of the higher interaction between the ester groups and dipolar sulfonyl groups. Chattopadhyay et al. [25] studied the effect of chain extender structure (BD, 1,2-propanediol, 2,2-diethyl-1,3-propanediol, 2-butyl 2-ethyl-1,3-propanediol, 4,4′-diaminodiphenyl sulfone and 4,4′-diaminodiphenyl methane) on the properties of polyurethanes. 4,4′-Diaminodiphenyl sulfone-based polyurethane exhibited higher tensile strength, storage modulus, elongation at break as compared to the other polyurethanes. This could be caused by the polar nature of sulfone group, which acts as a pseudo-cross-linker and increases the phase separation.

Chun et al. [26] studied the effect of chain extender, i.e., BD and ethylenediamine (ED) on the shape memory properties of shape memory polyurethanes. Both the polyurethanes exhibited almost similar shape recovery ratio. However, the ED-based polyurethane showed higher shape retention behavior. This is due to the presence of urea bonding, which restricted the chain rotation or slippage resulting in the rigid structure. Kalajahi et al. [27] also demonstrated the effect of chain extenders on the shape memory properties of polyurethanes. They used four different types of chain extenders such as ethylenediamine, 1,4-diaminobutane and 1,6-diaminohexane as aliphatic chain extenders and piperazine as cyclo-aliphatic chain extender. The cyclo-aliphatic chain extender-based polyurethane exhibited better shape fixity ratio as compared to the aliphatic chain extender-based polyurethanes. This is because of the high incompatibility between hard segments and PCL soft segments due to the presence of stiff cyclo-aliphatic piperazine chain extender. Further, the shape fixity decreased with the increase in carbon chain length of the aliphatic chain extenders. This was attributed to the increased thermodynamic compatibility between the hard and soft segments result in the decrease in crystallinity of soft segments by increasing chain extender length.

The effect of hard segment architecture on the SME of shape memory polyurethanes has been investigated by Wu et al. [28]. They designed the polyurethanes using PCL as soft segment, MDI and azobenzene diol or azobenzene diamine as chain extender, yielding shape memory polyurethanes containing azobenzene in side chains and in main chains. Polyurethanes with azobenzene in main chain exhibited the higher shape recovery ratio as compared to the polyurethane with side-chain azobenzene. This is due to the stronger interactions such as π–π interaction and induced dipole–dipole interaction between aromatic rings in the hard segments as a result of azobenzene in main chain of the polyurethane. T_m, i.e., T_{trans} of the polyurethane with side-chain azobenzene found to decrease when compared to polyurethane in the main-chain azobenzene. Azobenzenes in the side chains as hard segments increase the sequence randomness of the macromolecular chains and also restrict the segmental mobility, as a consequence decrease in T_m of PCL crystallization. Tsai et al. [29] investigated the effect of linear and dendritic side-chain extenders on the shape memory behaviors of polyurethanes. The linear polyurethane displayed good shape fixity, but quite poor shape recovery. The poor shape recovery is because of the presence of phase-segregated morphology and the lack of hydrogen-bonding interactions. The dendritic side-chain extender-based polyurethane showed better shape recovery behavior as compared to linear one. This is due to the peripheral alkyl chains of dendrons that promote the miscibility between hard segments and soft segments consequently inducing phase mixing morphology as well as rich hydrogen-bonding interactions that reinforce the physical cross-linking effect of hard segments.

4.2.4 Catalyst

Catalysts are added to enhance the reaction rates. Reaction occurs at lower temperatures, deblocking the blocked isocyanates and decreasing the curing temperature and time. The mostly used catalysts are aliphatic and aromatic amines (1,4-diazabicyclo[2.2.2]octane [DABCO], 1,8-diazabicyclo[5.4.0]undec-7-ene [DBU], 2,2′-bis-(dimethylaminoethyl ether) and triethylamine), organic acids (triflic acid, trifluoromethanesulfonimide, *p*-toluenesulfonic acid and methanesulfonic acid), organometallic compounds (dibutyltin diacetate, dibutyltin dilaurate [DBTDL], stannous octoate and iron acetylacetonate) and guanidines (naphthyl bis-guanidine and *N*-methyl-1,5,7-triazabicyclododecene [MTBD]) [2, 30]. Among the amines, tertiary amines are widely used in urethane formation reactions. Catalytic action occurs through the complex formation between isocyanate and the tertiary amine by donating the electrons on nitrogen atom of tertiary amine to the positively charged carbon atom of the isocyanate, followed by the nucleophilic attack of the alcohol. The conversion of the isocyanate alcohol reaction depends on the basicity and concentration of the amines. The increased basicity enhances the catalytic activity of the isocyanate alcohol reaction. In the case of acid catalyst, a dual hydrogen-bonding mechanism involves followed by electrophilic activation of the isocyanate via the isocyanate nitrogen, with simultaneous nucleophilic activation of the alcohol. Metal-based catalysts serve as Lewis acids toward alcohol or isocyanate to initiate the polymerization. The positive metal center interacts with electron-rich oxygen atom of both the hydroxyl and isocyanate groups that produce an intermediate complex, which in turn rearranges to form the urethane bond. Organotin compounds have predominantly served the active catalytic role; however, their removal from polyurethanes is often very difficult, resulting in detrimental effects on the aging of the final material. Recently, strong efforts have been given to catalyze the polyurethane reactions using organic acids, guanidines and heterocyclic carbenes. Catalysts used in polyurethane synthesis are presented in Table 4.4.

The effect of sequential addition of DBTDL catalyst on the reaction rate of polyurethane formation was studied [31]. It was found that the catalytic activity depended on the sequential addition of DBTDL catalyst. It was observed that the reaction rate was slower when the catalyst was added to diol followed by the addition of diisocyanate when compared to the catalyst added to the mixture of diol and diisocyanate. The effect of different catalysts (tin octoate, DBTDL, iron acetylacetonate and copper acetylacetonate) on the reaction rate and the properties of polyurethanes have been

Table 4.4: Catalysts used for the synthesis of shape memory polyurethanes.

Name of catalyst	**Chemical structure**
DABCO	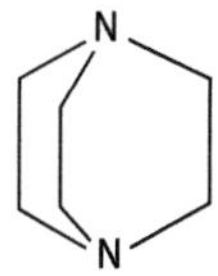
DBU	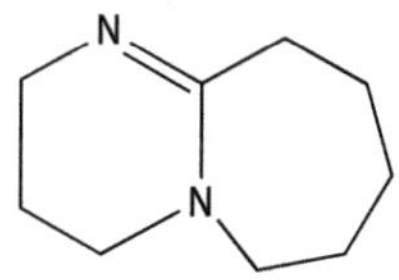
Triflic acid	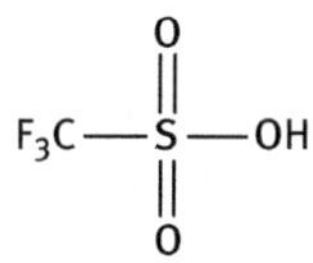
p-Toluenesulfonic acid	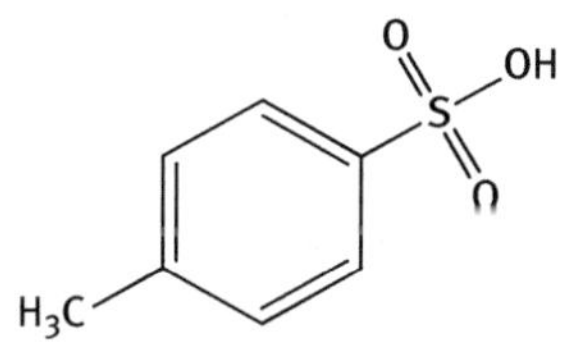
Trifluoromethanesulfonimide	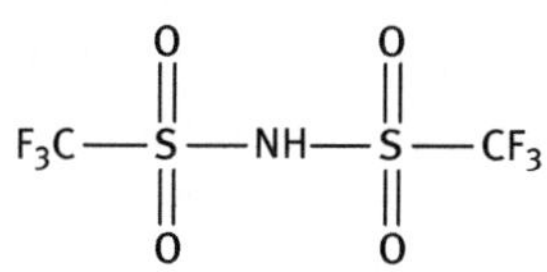
DBTDL	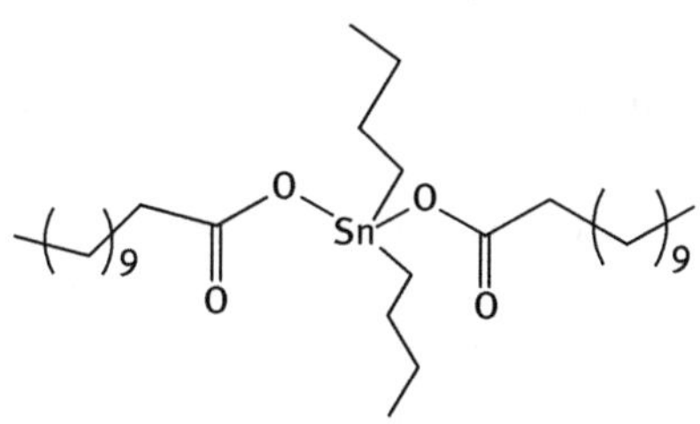
MTBD	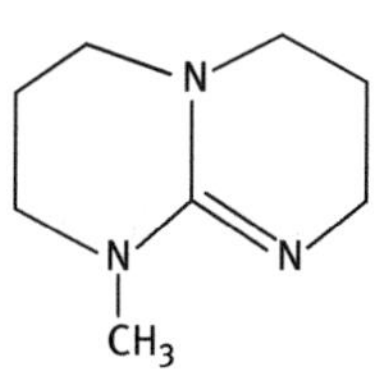

DABCO, 1,4-diazabicyclo[2.2.2]octane; DBU, 1,8-diazabicyclo[5.4.0]undec-7-ene; DBTDL, dibutyltin dilaurate; MTBD, *N*-methyl-1,5,7-triazabicyclododecene.

reported [32]. The DBTDL and iron acetylacetonate catalyzed reaction showed the higher reaction rate when compared to the other catalysts. The catalysts studied can be arranged in the order of decreasing activity as DBTDL > iron acetylacetonate > tin octoate > copper acetylacetonate. It was reported that the strength and thermal stability of polyurethanes catalyzed by DBTDL and iron acetylacetonate will be decreased. This is due to the accelerating catalytic effect of these compounds on the thermo-oxidative destruction processes occurring in curing of molded formulation under thermostating conditions. Sardon et al. [33] studied the activity of acid organocatalysts on the diisocyanate and polyol reaction along with the properties of polyurethanes. They have used various acid catalysts such as triflic acid, methanesulfonic acid, *p*-toluene sulfonic acid, diphenyl phosphate, trifluoroacetic acid and acetic acid. They have demonstrated that sulfonic acids were found to be most effective for urethane formations among organic acids and even as compared to conventional tin-based catalysts such as DBTDL and DBU. However, phosphonic and carboxylic acids showed considerably lower catalytic activities. Furthermore, sulfonic acids conferred polyurethanes with higher molecular weights when compared to traditional tin-based catalyst. It was suggested that the strength of a given acid as well as the nucleophilicity of its conjugate base was crucial in catalyzing the urethane formation. Alsarraf et al. [34] reported that guanidines are efficient organocatalysts for the synthesis of polyurethanes. It was observed that bicyclic penta-alkylated guanidines such as MTBD led to polyurethane molecular weight and dispersity in the range of those observed with tin-based catalyst such as DBTDL. Tetra-alkylated guanidine such as TBD exhibited weaker catalyst as compared to penta-alkylated guanidines due to its high reactivity toward isocyanate, as a consequence in the formation of a less nucleophilic urea. Coutelier et al. [35] reported the *N*-heterocyclic carbine (NHC) catalyzed synthesis of polyurethanes. They have demonstrated that NHC can efficiently catalyze the metal-free step-growth polymerization reaction of aliphatic diisocyanates with polyols, forming soluble linear polyurethanes. The NHC catalyst found to be much more efficient than a tertiary amine-based catalyst such as DABCO, enabling a polymerization at a relatively low catalyst concentration (1%) and low temperatures (30–50 °C). It was proposed that the urethane bond formation by NHC catalysis takes place by a basic deprotonation of alcohol by NHC, followed by a nucleophilic attack of the activated alcohol onto the isocyanate moiety.

4.3 Synthesis of shape memory polyurethanes

4.3.1 One-shot method

One-shot and pre-polymerization methods are used for the synthesis of polyurethanes. In the one-shot method all the reactants namely macroglycol, diisocyanate

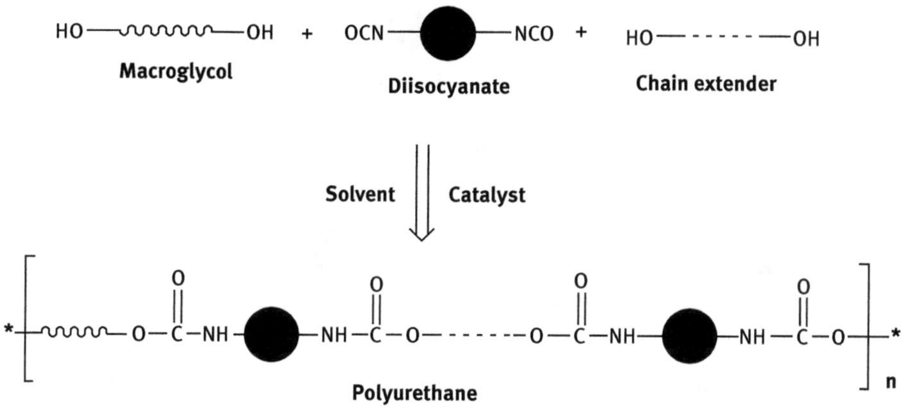

Figure 4.1: Schematic representation of one-shot method for polyurethane synthesis.

and chain extender are added at once at the starting of the polymerization reaction (Figure 4.1). This process is easier, faster and more reproducible. A high order of crystallinity is obtained in this process [10]. However, controlling of reaction is difficult. A wider distribution of hard segment lengths is obtained as all the reactants are fed together which lead to compete with the oligomeric diol and short-chain diol to react with diisocyanate, resulting in some longer hard segments. The reactivity of the polyol is an important factor in determining whether or not a polyol can be used in a one-shot technique. As a general rule, the reactivity of polyol needs to be nearly equal to or greater than the reactivity of the curative. For example, most of the MDI-based polyurethane elastomers in one-shot process use BD as a curative which contains primary hydroxyl groups. Castor oil contains secondary hydroxyl groups. In the reaction of MDI, BD and castor oil, the MDI will preferentially react with the BD to form MDI-BD hard segment. The MDI-BD hard segment will be phase separated before reacting with the castor oil because of the incompatibility in the system. This renders the poor physical properties as a result of the low-molecular-weight polymer. In a consequence, the polyol and curative should have similar reactivity for successful one-shot technique.

4.3.2 Pre-polymerization method

In the pre-polymerization method, –NCO- or –OH-terminated pre-polymer is prepared by reacting a diisocyanate with a macroglycol in the first step. Then a chain extender is added to the pre-polymer in the second step and high-molecular-weight polyurethane is obtained (Figure 4.2). This method is found to be more controllable. More regular hard-soft-hard-soft sequences and narrow distribution of hard segment lengths are obtained in this technique [10, 36]. This is because the soft segment is formed first by reacting the diisocyanate with oligomeric diol, then hard segments

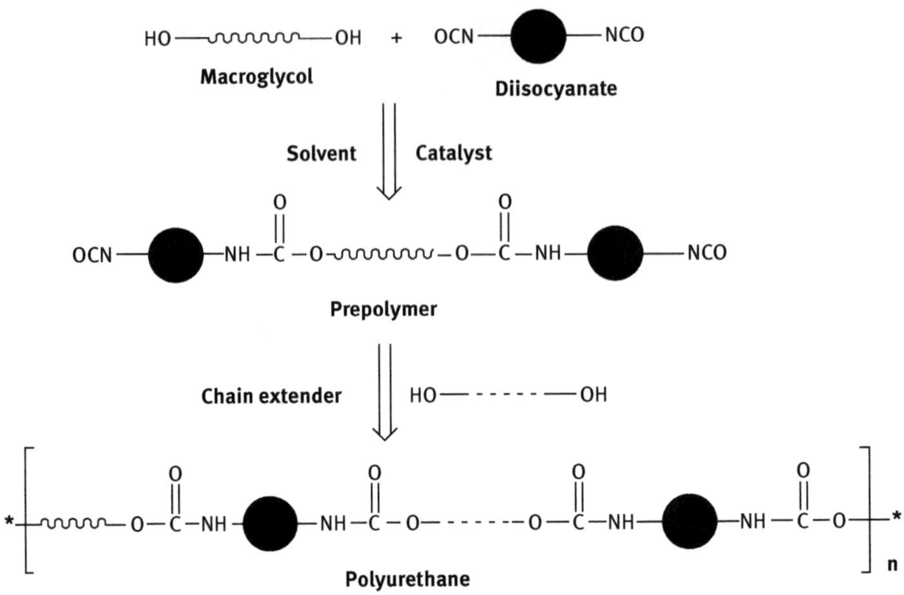

Figure 4.2: Schematic representation of pre-polymerization method for polyurethane synthesis.

are formed after adding the short-chain diol. The structural regularity and narrow hard segment distribution may result in the higher mechanical properties as well as better SME. Peebles theoretically demonstrated that under ideal conditions and at complete conversion under stoichiometric conditions the sequence length distribution of hard block segments in a segmented block copolymer follows the most probable distribution. A two-stage polymerization results in a narrower distribution of hard blocks than a single-stage polymerization of the same stoichiometric condition when the first reaction of the low-molecular-weight difunctional monomer is faster than the second reaction. This result was due to the factors of steric hindrance, alteration of the induction–resonance condition of the molecule when it is partly reacted, or the molecular configuration in the transition state.

The effect of polymerization methods, i.e., one-shot and pre-polymerization on physical properties of thermoplastic polyurethane elastomers was studied by Ahn et al. [37]. It was observed that a broader distribution of hard segment lengths was obtained prepared by the one-shot method, whereas narrow distribution was obtained in case of pre-polymerization method. These evidences were clearly observed in the differential scanning calorimetry analysis. The melt crystallization temperature of the hard segment phase prepared by pre-polymerization technique showed a sharp exotherm. However, a broader exotherm at higher temperature was obtained prepared by one-shot technique. Polyurethane prepared by one-shot method showed higher tensile strength and elongation at break when compared to polyurethane prepared by pre-polymerization method. These may be due to the well-developed hard segment domains caused by longer hard segments that serve as physical cross-links or fillers.

References

[1] Park JH, Kim BK. Infrared light actuated shape memory effects in crystalline polyurethane/graphene chemical hybrids. Smart Mater Struct 2014;23:1–7.

[2] Sharmin E, Zafar F. Polyurethane: an introduction. London: INTECH Open Access Publisher, 2012:3–16.

[3] Kalita H, Karak N, Mesua ferrea L. seed oil-based hyperbranched shape memory polyurethanes: effect of multifunctional component. Polym Eng Sci 2012;52(11):2454–61.

[4] Kalita H, Karak N. Bio-based elastomeric hyperbranched polyurethanes for shape memory application. Iran Polym J 2012;21(4):263–71.

[5] Kalita H, Karak N. Biobased hyperbranched shape-memory polyurethanes: effect of different vegetable oils. J Appl Polym Sci 2014;131(1):1–8.

[6] Mo F, Zhou F, Chen S, Yang H, Ge Z, Chen S. Development of shape memory polyurethane based on polyethylene glycol and liquefied 4,4'-diphenylmethane diisocyanate using a bulk method for biomedical applications. Polym Int 2015;64(4):477–85.

[7] Wang HH, Yuen UE. Synthesis of thermoplastic polyurethane and its physical and shape memory properties. J Appl Polym Sci 2006;102(1):607–15.

[8] Lee BS, Chun BC, Chung Y-C, Sul K II, Cho JW. Structure and thermomechanical properties of polyurethane block copolymers with shape memory effect. Macromolecules 2001;34(18):6431–7.

[9] Chun BC, Cho TK, Chung YC. Blocking of soft segments with different chain lengths and its impact on the shape memory property of polyurethane copolymer. J Appl Polym Sci 2007;103: 1435–41.

[10] Prisacariu C. Polyurethane elastomers: from morphology to mechanical aspects. New York: Springer, 2011.

[11] Javni I, Zhang W, Petrović ZS. Effect of different isocyanates on the properties of soy-based polyurethanes. J Appl Polym Sci 2003;88(13):2912–16.

[12] Desai S, Thakore IM, Sarawade BD, Devi S. Effect of polyols and diisocyanates on thermo-mechanical and morphological properties of polyurethanes. Eur Polym J 2000;36(4):711–25.

[13] Pandya MV, Deshpande DD, Hundiwale DG. Effect of diisocyanate structure on viscoelastic, thermal, mechanical and electrical properties of cast polyurethanes. J Appl Polym Sci 1986;32(5):4959–69.

[14] Yang JH, Chun BC, Chung YC, Cho JH. Comparison of thermal/mechanical properties and shape memory effect of polyurethane block-copolymers with planar or bent shape of hard segment. Polymer 2003;44(11):3251–8.

[15] Ping P, Wang W, Chen X, Jing X. The influence of hard-segments on two-phase structure and shape memory properties of PCL-based segmented polyurethanes. J Polym Sci Part B Polym Phys 2007;45(5):557–70.

[16] Ji FL, Hu JL, Li TC, Wong YW. Morphology and shape memory effect of segmented polyurethanes. Part I: with crystalline reversible phase. Polymer 2007;48(17):5133–45.

[17] Ionescu M. Chemistry and technology of polyols for polyurethanes. Ohio: Rapra, 2005.

[18] Martin DJ, Meijs GF, Renwick GM, Gunatillake PA, Mccarthy SJ. Effect of soft-segment CH2/O ratio on morphology and properties of a series of polyurethane elastomers. J Appl Polym Sci 1996;60(4):557–71.

[19] Kim BK, Lee SY, Xu M. Polyurethane having shape memory effect. Polymer 1996;37(26):5781–93.

[20] Li F, Zhang X, Hou J, Xu M, Luo X, Ma D, et al. Studies on thermally stimulated shape memory effect of segmented polyurethanes. J Appl Polym Sci 1997;64(8):1511–16.

[21] Ahmad M, Xu B, Purnawali H, Fu Y, Huang W, Miraftab M, et al. High performance shape memory polyurethane synthesized with high molecular weight polyol as the soft segment. Appl Sci 2012;2(4):535–48.

[22] Cho JW, Jung YC, Chung YC, Chun BC. Improved mechanical properties of shape-memory polyurethane block copolymers through the control of the soft-segment arrangement. J Appl Polym Sci 2004;93(5):2410–15.
[23] Zhu Y, Hu J, Yeung K. Effect of soft segment crystallization and hard segment physical crosslink on shape memory function in antibacterial segmented polyurethane ionomers. Acta Biomater 2009;5(9):3346–57.
[24] Liaw D. The relative physical and thermal properties of polyurethane elastomers: effect of chain extenders of bisphenols, diisocyanate, and polyol structures. J Appl Polym Sci 1997;66(7):1251–65.
[25] Chattopadhyay DK, Sreedhar B, Raju KV. Effect of chain extender on phase mixing and coating properties of polyurethane ureas. Ind Eng Chem Res 2005;44(6):1772–9.
[26] Chun BC, Cho TK, Chung YC. Enhanced mechanical and shape memory properties of polyurethane block copolymers chain-extended by ethylene diamine. Eur Polym J 2006;42(12):3367–73.
[27] Kalajahi AE, Rezaei M, Abbasi F, Mir Mohamad Sadeghi G. The effect of chain extender type on the physical, mechanical, and shape memory properties of poly(ε-Caprolactone)-based polyurethane-ureas. Polym Plast Technol Eng 2017;2559(March):1–9.
[28] Wu X, Liu L, Fang W, Qiao C, Li T. Effect of hard segment architecture on shape memory properties of polycaprolactone-based polyurethane containing azobenzene. J Mater Sci 2016;51(5):2727–38.
[29] Tsai CC, Chang CC, Yu CS, Dai S, Wu TM, Su WC, et al. Side chain dendritic polyurethanes with shape-memory effect. J Mater Chem 2009;19(44):8484.
[30] Sardon H, Pascual A, Mecerreyes D, Taton D, Cramail H, Hedrick JL. Synthesis of polyurethanes using organocatalysis: a perspective. Macromolecules 2015;48(10):3153–65.
[31] Niyogi S, Sarkar S, Adhikari B. Catalytic activity of DBTDL in polyurethane formation. Indian J Chem Technol 2002;9(4):330–3.
[32] Bakirova IN, Kirillova AS. Effect of organometallic catalysts on the synthesis process and properties of molded polyurethane. Russ J Appl Chem 2013;86(9):1399–403.
[33] Sardon H, Engler AC, Chan JMW, Garcia JM, Coady DJ, Pascual A, et al. Organic acid-catalyzed polyurethane formation via a dual-activated mechanism: unexpected preference of n-activation over o-activation of isocyanates. J Am Chem Soc 2013;135(43):16235–41.
[34] Alsarraf J, Ammar YA, Robert F, Cloutet E, Cramail H, Landais Y. Cyclic guanidines as efficient organocatalysts for the synthesis of polyurethanes. Macromolecules 2012;45(5):2249–56.
[35] Coutelier O, El Ezzi M, Destarac M, Bonnette F, Kato T, Baceiredo A, et al. N-Heterocyclic carbene-catalysed synthesis of polyurethanes. Polym Chem 2012;3(3):605.
[36] Peebles LH. Hard block length distribution in segmented block copolymers. Macromolecules 1976;9(1):58–61.
[37] Ahn TO, Choi IS, Jeong HM, Cho K. Thermal and mechanical properties of thermoplastic polyurethane elastomers from different polymerization methods. Polym Int 1993;31(4):329–33.

5 Applications of shape memory polymers

5.1 Introduction

Shape memory polymers (SMPs) have promising applications in the field of sensors and actuators, aerospace engineering, textile engineering, automobile engineering, packaging, self-healing, self-peeling reversible adhesive and biomedical field [1–5]. SMPs are being widely developed and qualified for deployable components and structures in aerospace. Composite Technology Development Inc., USA, designed and developed a simple and inexpensive deployable solid-surface reflector using TEMBO® SMP composite. In the textile and apparel applications, SMPs are gaining much attention to develop an intelligent waterproof and breathable fabrics. Mitsubishi Heavy Industries, Japan, successfully developed Diaplex™ smart fabrics using SMP with excellent waterproofing and breathability. Wrinkle-free and crease retention SMP finishing fabrics have been developed. Self-healing SMP can play an important role in the recovery of surface damages/cracks during the damaging and scratching of a surface. SMPs have broad applications in biology and medicine, especially for biomedical devices, which might open new medical procedure for minimally invasive surgery. The bulky SMP device can be inserted into the body in a small fixed shape by a small laparoscopy and the permanent shape can be recovered under the application of suitable stimulus. Many research groups and industries are pursuing medical applications of SMP. However, limited numbers of biomedical SMP devices are commercially available due to the lack of supplier of medical grade materials and precision of SMPs. MedShape Inc., USA, has developed SMP suture (Morphix®). SMPs have fantastic development prospects and value in various important engineering and advanced technological fields.

5.2 Biomedical applications

SMPs have potential applications in the field of biomedicine [1, 4]. Biocompatible and biodegradable SMPs stimulate to develop implantable SMP devices. The actuation temperature of SMP can be modified by changing chemical composition that assists in developing self-actuated biomedical SMP devices, which can be actuated at body temperature. The easy fabrication of SMP into complex shapes takes the benefit to develop custom patient-specific device geometries. Biomedical SMP devices such as clot removal device, vascular stent, surgical suture and orthodontic appliance are being developed and are discussed below.

https://doi.org/10.1515/9783110570175-005

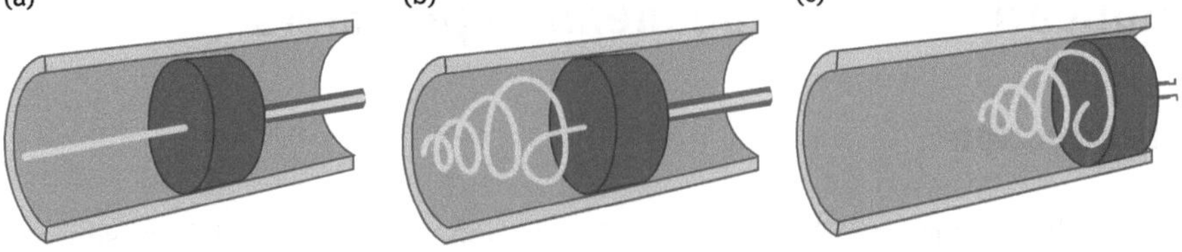

Figure 5.1: Depiction of endovascular clot removal using the SMP microactuator: (a) the device first punctured the clot, (b) activated to a form of coil shape on the distal side of the clot and (c) pulled to remove both the device and clot simultaneously.

5.2.1 Clot removal device

SMPs can be utilized for the treatment of ischemic stroke [6, 7]. Ischemic strokes are caused by the formation or lodging of a blood clot (thrombotic vascular occlusion) in the arterial network that deprived the brain of oxygen supply. This is a mechanical clot extraction device, as an alternative to conventional clot-dissolving drug treatment. These devices, i.e., SMP microactuators were designed to be delivered through a catheter and punctured the clot in a narrow form. Then actuated by external stimuli to deploy into a clot-grabbing form such as corkscrew or coil, remove both the microactuator and blood clot (thrombus) restoring the blood flow (Figure 5.1). Maitland et al. [6] reported the photothermal design aspect of novel SMP microactuators for stroke treatment. Thermoset shape memory polyurethanes were used to design the device. Diode laser operating at a wavelength of 810 nm was used to actuate the SMP device. The SMP was doped with a laser-absorbing dye to attain selective heating of the device. Small IV et al. [7] demonstrated the laser-activated SMP intravascular thrombectomy device to mechanically retrieve the thrombus. The device consisted of thermoplastic shape memory polyurethane microactuator coupled with optical fiber for delivery of 810 nm laser light for stimulation. The SMP microactuator was programmed in a straight rod shape and its primary shape was a tapered corkscrew. The SMP microactuator was delivered in its secondary straight rod form through a catheter distal to the vascular occlusion, which is located by fluoroscopy routinely used in stroke treatment. Then the microactuator was transformed into its primary corkscrew shape by laser heating. Once deployed, the microactuator was retracted and the captured thrombus is removed from the body to restore the blood flow. Small IV et al. [8] also developed prototype using SMP encapsulating shape memory alloy (SMA, nitinol) electromechanical hybrid microactuator for endovascular thrombectomy device. SMP was employed to maintain the straightening stress for endovascular delivery of the device as well as relaxing the stress to allow for actuation of the nitinol to primary shape (corkscrew). An electric current was supplied to the microactuator; the temperature of nitinol was increased by Joule heating effect and consequently heating of the surrounding SMP. SMP (T_g = 86 °C) was transformed to its low-modulus rubbery state and relaxed the loading stress, as a consequence permitted the nitinol to resume its corkscrew shape. Actuation was achieved within 5 s in the water-filled model with the applied current of 0.8 A. It was observed that higher current was needed to achieve the actuation in water as compared to air because heat loss is faster in water than in air due to the higher thermal conductivity of water. It was suggested that the use of SMP with low T_g in an optimized device would reduce the thermal damage on the artery and offer potential therapeutic alternative for acute ischemic stroke.

5.2.2 Vascular stent

A stent is a tube made up of metal or polymer inserted into an abnormally narrowed or closed artery or duct (restenosis) in the human body to open it mechanically and restore the blood flow inside it. Metallic stents made up of SMA are too stiff to navigate through tortuous vessels and arises a compliance mismatch with the arterial walls, which result in an abnormal stress concentration initiating an adaptive response in the vascular tissue. SMP-based stents can overcome these problems because of their good flexibility, biodegradability, enhanced compliance matching, minimally invasive benefits, high drug-loading capacity, and less thrombogenesis and intimal hyperplasia. Furthermore, SMP stents could be actuated gently at body temperature, which reduces the thermal damage of surrounding tissues. A variety of thermoplastic SMPs have been reported in stent applications. However, most of the reported SMP stents are based on thermoplastic shape memory polyurethanes. The performance of the SMP stent depends on T_g, cross-link density, geometry and programming conditions. Baer et al. [9] reported the fabrication and in vitro deployment of a laser-activated SMP vascular stent. The stent was made from thermoplastic polyurethane and the photothermal actuation of the stent was employed in a water-filled mock artery. The stent did not fully deploy at the maximum laser power (8.6 W) under physiological flow rate, due to convective cooling. However, under zero flow, simulating the technique of endovascular flow occlusion, full deployment was achieved in the mock artery at a laser power of ~8 W. In vivo survival studies are required to evaluate the extent and tolerance of potential thermal damage of surrounding tissues caused by photothermal actuation of the SMP stent. Kim et al. [10] reported the feasibility of the development of thermoresponsive braided stents made of shape memory polyurethane. The simulations in this study demonstrated the proper choosing of diameter of the constituent shape memory polyurethane fibers and braiding angles, and the SMP braided stent could behave mechanically like the SMA stent. The result showed that no sudden overpressure was observed on the vessel due to gradual deployment, thereby reducing the possibility of hurting the vessel wall.

Drug-eluting stents are the most promising in recent advancement of stent technology. Drugs (e.g., sirolimus, paclitaxel and dexamethasone) reduce the restenosis by inhibiting proliferation, migration and growth of smooth-muscle cells (SMC). As a consequence reduces the SMC hyperplasia that forms a layer on the intimal wall of the vessel. A drug-eluting SMP stent has been developed by Chen et al. [11]. A sirolimus-eluting biodegradable stent made from chitosan-based strips fixed by an epoxy compound coated with a hydrophobic heparin was demonstrated. Sirolimus drug prevents the restenosis by inhibiting the proliferation and migration of vascular SMC, and heparin prevents the blood coagulation around the stent. The stent showed the hemo- and cytocompatibility as observed in the 3-(4,5-dimethylthiazolyl-2)-2,5-diphenyltetrazolium bromide (MTT) assays and live/dead cell viability. The stent could potentially be inserted into an artery using minimally invasive surgery because it can self-expand from a crimped state to its expanded state stimulated by hydration. The animal study indicated that the stent can be reliably placed in arteries and effectively suppress the restenosis.

Biodegradable polymer stents are more promising because of their bioresorption, more drug-loading capacity, less risk of mechanical blockage and repeat surgery. However, some biodegradable polymer stents also have encountered some problems during biodegradation. Degradation of acid–base polymers leads to a significant pH change due to the release of acid, which results in the necrosis of the surrounding tissues. Most of the biodegradable polymers have low strength that may cause the collapse of stent before the complete degradation leads to the blockage of the vessel. Many researchers have focused on development of biocompatible, fully biodegradable and excellent mechanical strength SMP stent. Tamai et al. [12] developed biodegradable poly(l-lactic acid) (PLLA) coronary stents. They have evaluated the safety and efficacy of PLLA stent implanted in 15 patients. No stent thrombosis and no major cardiac event were observed within 30 days. All patients showed acceptable restenosis and target lesion revascularization rates in the 6-month period of test. Body temperature-responsive fully biodegradable polymeric stent based on PLLA and poly(glycolic acid) (PLGA) bilayers have been reported by Venkatraman et al. [13]. The single-layer stent made from PLLA displays very slow self-expansion at body temperature, whereas rapid expansion was observed by the addition of a layer of PLGA to the PLLA. It was observed that thickness, composition and T_g of each layer in the stents have most significant effect on the expansion rate. It was suggested that bilayered structures, with appropriate selection of the "top" and "bottom" layers, enable to develop rapid self-expandable stent at body temperature. Yang et al. [14] have developed biodegradable thermoresponsive PEG-PCL copolymer-based SMP stent. Cytotoxicity tests proved the good biocompatibility of the stent. In vitro degradation tests exhibited that stent experienced a bulk degradation of 47% after 60 days of incubation under flow conditions. The stent did not collapse even during the bulk degradation process, and this indicated the excellent mechanical strength of the stent.

5.2.3 Orthodontic appliance

Orthodontic treatment refers to the straightening or moving of misaligned teeth using orthodontic appliances commonly called braces. A brace consists of an orthodontic bracket and an orthodontic archwire. Nitinol (SMA) is often used as an orthodontic material because of its excellent performance. However, its metallic color renders aes-

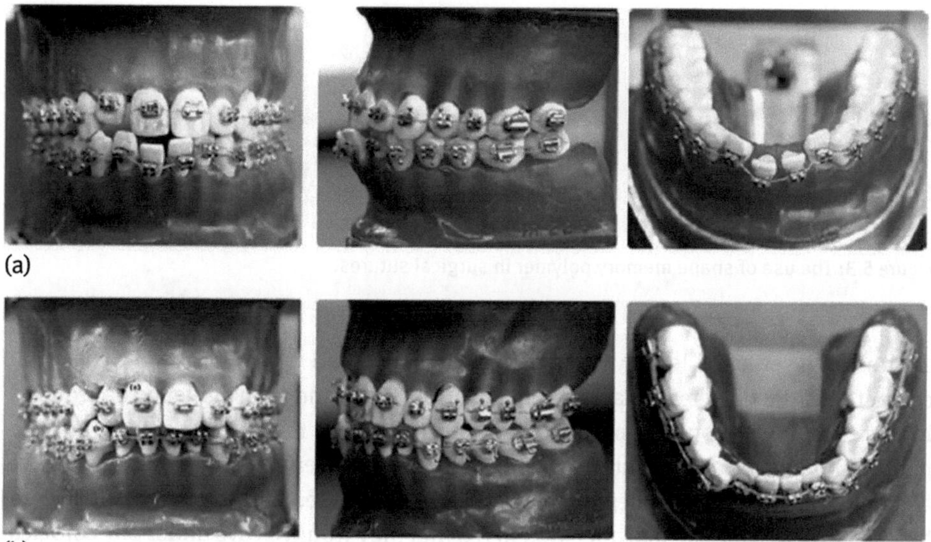

Figure 5.2: The utilization of shape memory wire in orthodontic application (a) before and (b) after orthodontic treatment.

thetic problems. SMP may be a good alternative to SMA because of high shape recovery, easy processing, transparency and satisfactory aesthetic appearance. Nakasima et al. [15] demonstrated the concept of using a thermoresponsive SMP archwire in orthodontic braces for aligning teeth. It was observed that force produced by SMP was sufficient for moving teeth and the SMP was more aesthetically appealing (nearly tooth colored) than a metallic archwire. Jung and Cho [16] reported the use of shape memory polyurethanes for orthodontic archwires. The orthodontic application test was carried out using polyurethane wire and was attached to a multibracket system on the dental model. Movement of the teeth was clearly observed when the appliance was heated above the T_{trans} (40 °C) and the misaligned teeth began to align due to the shape recovery of the wire between the brackets (Figure 5.2). It was indicated that the shape memory polyurethane would be realized in orthodontic treatment with satisfactory aesthetic appearance and constant recovery force over a long period of time.

5.2.4 Suture

SMP can be used as surgical suture, which will play a significant role in endoscopic surgery. Generally, sutures are to be stitched very carefully. They may damage the cells of the skin if the threads are stitched very tightly; on the other hand, if the threads are too loose, it will not do the function properly. Effective suturing depends on the skill and experience of the surgeon. The self-tightening SMP suture definitely overcomes these problems. SMP

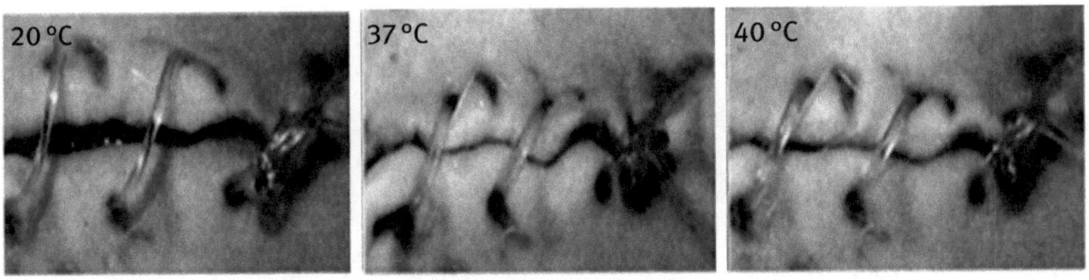

Figure 5.3: The use of shape memory polymer in surgical sutures.

suture can be stitched loosely in its temporary shape and when the sutures come into contact with the body temperature or T_{trans}, it will shrink and then tightened as required. In this case, it will apply an optimal force and reduce the risk of cell damage. Biocompatible and biodegradable sutures would be more prominent for wound closing device. Lendlein [17] developed biodegradable SMP surgical suture. They introduced hydrolyzable ester linkages into the polymer systems so that they would cleave under physiological conditions. The degradation kinetics can be controlled by varying the composition and relative mass content of the precursor macrodiols. The polymers show linear mass loss in vitro, resulting in a continuous release of degradation products leading to the less risk of inflammatory response. In the in vivo test it was observed that the knot was tightened and closes the wound when the temperature increased to 41 °C (Figure 5.3).

5.2.5 Dialysis needle

Vascular access complications occur in kidney dialysis patients as a result of arteriovenous (AV) graft failures. AV graft failure is caused by the development of intimal hyperplasia and the subsequent formation of stenotic lesions that decline graft flow. Vascular damage due to excessive wall shear stress may cause to lesion formation. SMP adapter for a kidney dialysis needle has been developed to decrease the hemodynamic stresses during dialysis [18]. An SMP adapter has been proposed to pass through the dialysis needle in a compact form, thermally expanded upon heating to body temperature and be retracted when the procedure is completed (Figure 5.4). Flow visualization experiments demonstrated that the graft wall shear stresses were significantly decreased by the adapter's elimination of jet impingement, which could potentially reduce vascular access occlusion.

5.3 Textile applications

Extensive efforts have been made in research and development to build up smart textile materials. SMP can be used in textile in various forms such as fibers, yarn and

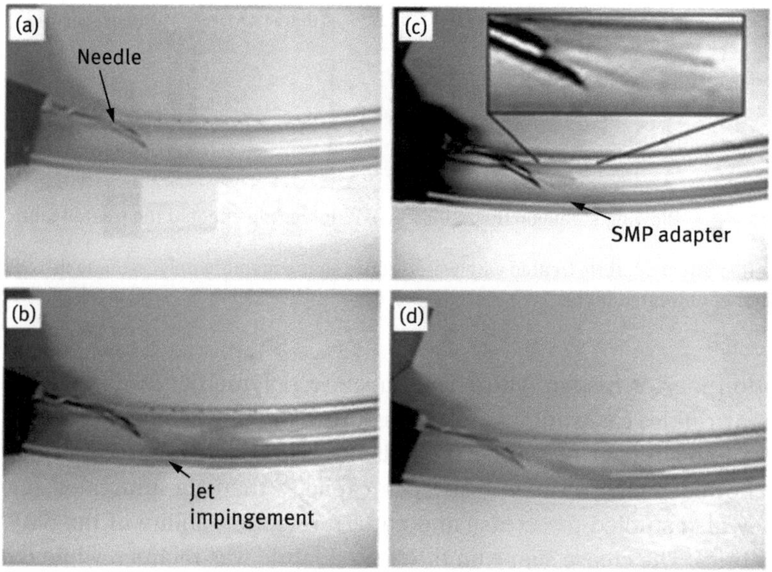

Figure 5.4: The use of shape memory polymer in dialysis needle: (a) a dialysis needle delivery, (b) flow visualization within the AV graft model, (c) deployed SMP adapter following delivery through the dialysis needle and thermal actuation and (d) visualization of the SMP adapter flow.

fabrics. SMPs bring tremendous interest for use in textile and clothing because they can respond to changes in heat and moisture levels, providing greater comfort for the wearer [2, 19–21]. The SMP fabric can be used in cuffs and collars, which need to keep their shape, and for knees and elbows, which need to recover their shape if wrinkled. SMP-coated or -laminated garments can control the water vapor diffusion and provide better comfort in both cold and warm climates. SMP-based garments can be deformed to different sizes as required to wearer's figure. SMP-coated or -laminated fabrics have extensive application in sportswear, footwear, gloves and socks because of their excellent waterproofing and breathability. SMPs also deliver some smart attributes, for example, aesthetic appeal, textile soft display, smart wetting properties, health monitoring and protection against extreme variations in environmental conditions. SMPs used in various ways in textile applications are discussed below.

5.3.1 Finishing fabrics

Thermoresponsive SMP can be used in the textile via garment finishing. SMP finishing fabrics deliver some properties such as wrinkle-free, crease and pattern retention, aesthetic appeal, antishrinkage and antifelting. The wrinkle-free and crease retention of the SMP finishing fabrics are due to the excellent shape recovery and shape fixity of the SMP, respectively. High shape recovery, high shape fixity and high elasticity of SMP are most important to achieve these properties. Hu et al. [22] reported the

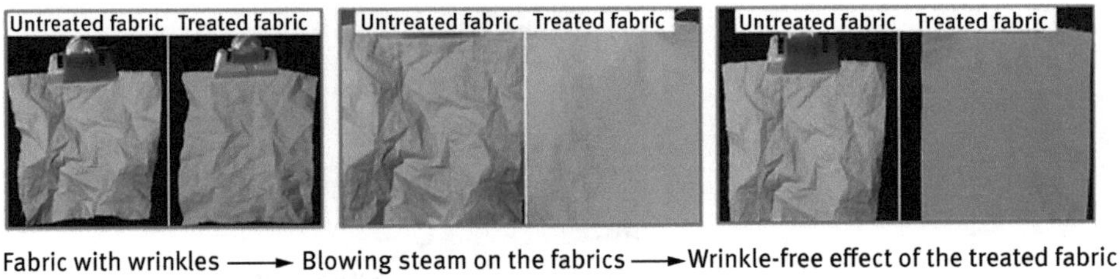

Figure 5.5: Wrinkle-free effect of fabric treated with water-borne shape memory polyurethane in comparison with that of untreated fabric.

wrinkle-free cotton fabrics treated with shape memory polyurethanes [22]. Cotton fabrics encounter wrinkles easily under low stress during wearing or storage because of the debonding and slippage of hydrogen bonds. It was observed that the treated cotton fabrics showed higher wrinkle resistance capacity than the untreated one (Figure 5.5). They also studied the crease and pattern retention ability of the SMP-treated cotton fabrics. The crease shape on the treated fabric was retained while the crease shape on the untreated fabric was disappeared after hot water washing. The treated cotton fabrics also showed the good antishrinkage ability by reducing the wool directional frictional effect. They have systematically investigated the use of shape memory polyurethanes in fabric finishing, and garment finishing lasts a decade. They also developed aqueous shape memory polyurethanes finishing agent to get excellent flexible, tactility and washable textiles.

5.3.2 Breathable fabrics

Breathable fabrics guard the human body from external heat, wind and many harmful agents, and simultaneously it also allows effective transmission of water vapor through the fabrics. Breathable fabrics passively permit water vapor to diffuse through them yet still prevent the penetration of liquid water. Breathable fabrics are developed by using the SMP. SMP used for breathable fabrics have a T_g at around body temperature. The water vapor permeability (WVP) of the SMP is changed according to the wearer's body temperature. By coating with SMP, the breathability of the garment can be controlled to a large extent. SMP-coated fabric has high WVP at higher temperature (above T_g) and keeps the body cool, and low WVP at low temperature keeps the body warm. The molecular free volume of the fabric increases significantly when the body temperature is above T_g of fabric and allows the transfer of heat and vapor to the environment resulting in a comfortable feeling. Shape memory polyurethanes show higher WVP at higher humidity and lower WVP at lower humidity. The degree of WVP of SMP can be tuned by changing the reacting materials and compositions. Ding et al. [23] studied the free volume and water vapor transport properties of temperature-responsive segmented polyurethanes [23]. It was observed that the mean free volume size and fractional free volume increased more rapidly in the temperature range of crystal melting than in other temperature intervals. WVP of the polyester-based polyurethane membranes increased significantly in the crystal melting temperature range. It was found that WVP of the polyester-based polyurethanes showed approximately direct correlation with the fractional free volume within the temperature range of crystal melting. Jeong et al. [24] have also investigated the WVP of shape memory polyurethanes with amorphous reversible phase. They modified the polyurethanes using hydrophilic segments, i.e., diol-terminated poly(ethylene oxide) (PEG 200) or dimethylolpropionic acid (DMPA). It was observed that the higher content of PEG 200-based polyurethanes showed higher WVP. The neutralized DMPA unit enhanced the WVP at the temperature range above T_g. WVP increased with the increase of PEG content due to the increase in the number of polar groups in the polymer backbone.

5.3.3 Damping fabrics

SMP can absorb impact energy due to their good damping properties at around T_g. AlliedSignal Inc. manufactured SMP fiber (Securus fibers)-based automotive seatbelt, which can effectively improve a passenger's safety by utilizing the damping effect of the SMP [25]. The Securus fibers are melt spun from shape memory poly(ethylene terephthalate)-poly(caprolactone) block copolymers. It is reported that Securus fiber can absorb energy from the body's forward motion as a result improving the safety of passengers during a crash.

5.3.4 Others

SMP fabric can be used in the wound dressing materials. SMP fabrics with noncytotoxic, hemolytic, sensitive or irritant properties are more pronounced for wound dressing applications. SMP fabric changes modulus as a response to body temperature change and thus the pressure applied on the wound may be tuned and a low pressure can be applied. Meng et al. [26] developed shape memory polyurethane fibers by spinning process and suggested that it could be good fabric for wound dressing. The fabric showed good cytocompatibility, nonhemolytic and nondermal irritant.

SMP can be used in wound healing monitoring textile. The pH-responsive SMP can provide information on the stage of the wound healing process. SMP can be utilized for preparing deodorant fabrics which can able to release deodorant agents at certain temperatures [27]. SMP fabric can be utilized as temperature and position sensors, and structural failure detectors.

5.4 Aerospace applications

SMAs are commercially available and found a wide range of applications in aerospace. However it has some disadvantages, for example, high weight, large volume, expensive and high deployable shock effect. SMPs have been gaining advantages in aerospace applications, because of their lightweight and variable properties above and below T_{trans}. SMPs are used in the deployment components and structures in aerospace [3, 28, 29]. Deployment structures are widely used in spacecraft to keep the large device in compact form in launch vehicle. The applications include solar arrays, solar panels, radar, antennas, hinges, truss and morphing skin,. all are deployable components. The deployable materials should possess light weight as possible, excellent strength to support structural load and resist external disturbances. Moreover, pure SMPs are less applicable in deployable structure because of their low strength and elastic modulus. Thus, SMP composites filled with carbon nanotube, carbon fiber and so on are usually used for the deployable structures. However, due to the extremely harsh space environment, so many important factors must be considered to use in structures of aerospace application.

5.4.1 Solar arrays

Solar arrays are the source of energy generation subsystem in space-deployable structure and packaged in satellite vehicle before launch. The solar arrays are released to deploy once reached in orbit and activated to generate energy. The work efficiency depends on the large area of deployable structure, toughness and light weight. Lan et al. [30] have developed carbon fiber-based thermoset styrene SMP composite hinge to deploy solar array. A prototype of a solar array of such an SMP composite hinge deploys successfully in 80 s under the application of 20 V. Composite Technology Development (CTD) and DR Technologies jointly have developed a deployable solar array design with a low cost, reliable and minimal overall part count using TEMBO® SMP composite hinges (Figure 5.6) [31]. The TEMBO® hinges are designed to drive and damp the deployment of the solar array and to lock the array after completion of deployment with no dead band.

5.4.2 Truss

SMPs can be used to actuate the inflatable truss from high packaging state to a large deployment state. Self-deploying SMP trusses have several advantages such as simple, low weight and inexpensive alternative to articulated mechanisms or inflatable structures. Important design parameters for a deployable truss are bending,

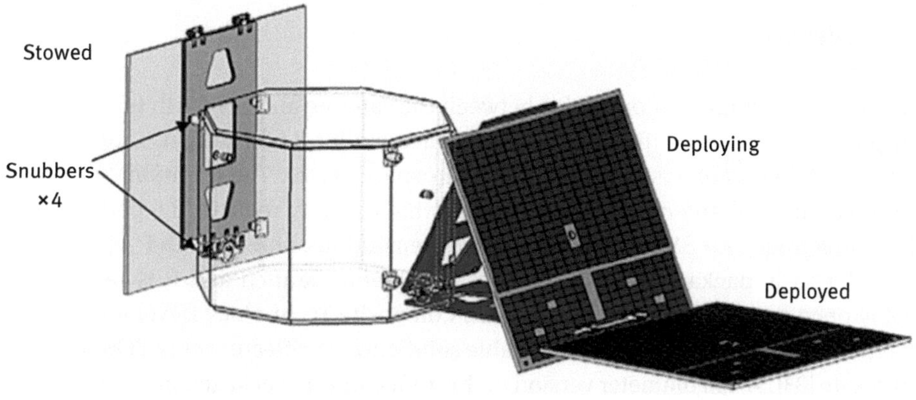

Figure 5.6: Deployable solar array designed by CTD.

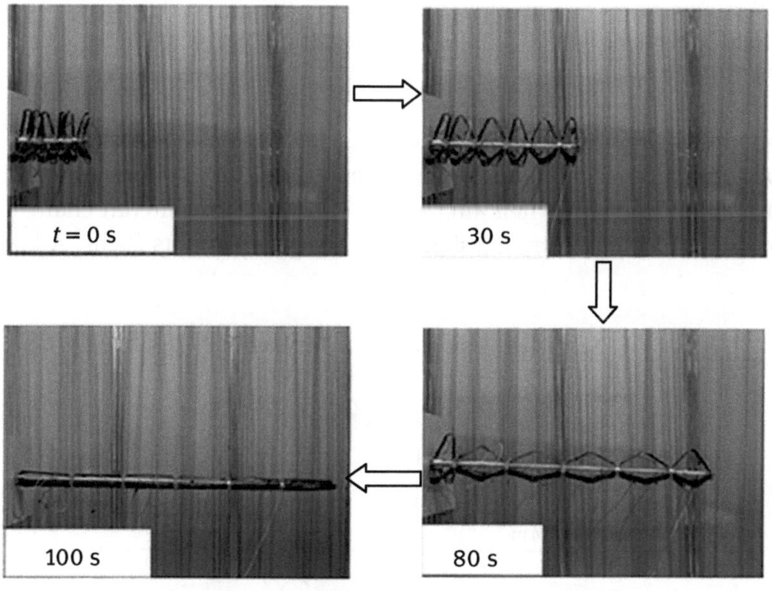

Figure 5.7: The unfold process of deployable SMP truss.

torsional and compression stiffness, stability and deployment repeatability. Zhang et al. [32] reported the theoretical analyses and experiments of a space-deployable truss structure made up of carbon fiber-reinforced epoxy-based SMP composites. The designed deployment truss completely unfolded within 100 s (Figure 5.7). The recovery moment at T_g was found to be about 0.29 N m. It was suggested that the fabricated deployable truss can be used as a deployable structure for aerospace applications and can provide the recovery moment for the deployable antenna and solar panel during their process of expansion effectively.

5.4.3 Antennas

Antenna is a communication device between the satellite and the Earth in space that can provide necessary information about space matter. Light weight, large diameter deployable antenna is required in the aerospace sector. The traditional mechanical drive mode of antenna has drawbacks, which greatly reduces the performance of the antenna. The SMP composites have tremendous potential and advantages. The antenna is packaged into a compact form before launch and it is released to deploy once on orbit by external stimulus. Composite Technology Development Inc. designed a simple and low-cost deployable solid-surface reflector using TEMBO® SMP composite [33]. A 4 m diameter version of this reflector was packaged in 1.4 m Falcon 1 e launch vehicle fairing. The designed compact antenna fully deployed upon heating (Figure 5.8).

5.4.4 Morphing structure

Aircrafts are often envisioned to be multifunctional for dynamic performance so that they can perform multiple missions during a single flight. The performance and efficiency of aircraft deteriorated when aircraft turns. A morphing aircraft changes its geometry to adapt to a changing mission environment during its flight. A key

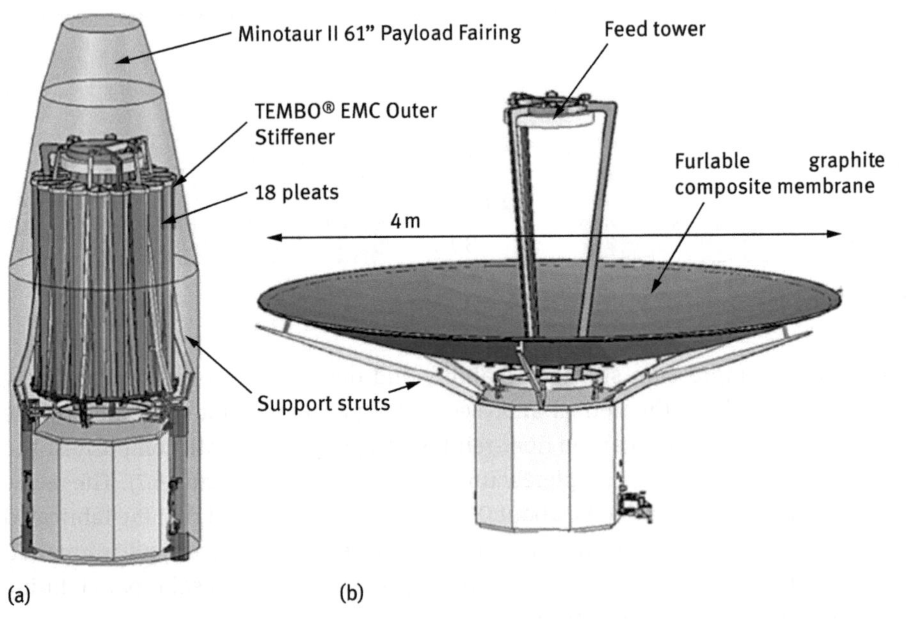

Figure 5.8: Solid-surface reflector developed by CTD: (a) the stowed state and (b) the deployed state.

technology to facilitate morphing aircraft is flexible skins. The skin should be able to withstand the out-of-plane aerodynamic loads while simultaneously carrying some shear loading. Skin material of the morphing wing should be flexible, elastic, high stain recovery, excellent abrasion resistant, excellent different weather resistant and hard enough to withstand the aerodynamic loads of the aircraft in different flight modes [34, 35]. SMPs are promising materials due to the literate stiffness and geometric changes at temperature changes. The feasibility of SMP morphing skin materials was demonstrated by Lockheed Martin Aeronautics Company. Yu et al. [36] designed a morphing wing based on SMP composite. The designed SMP composite morphing wing showed efficient deployment and wind-resistant ability. Yin et al. [34] developed the method of analyzing the deformation of flexible SMP skins under airflow. They investigated the out-of-plane deformation of the SMP skin at different temperatures. It was observed that the maximum out-of-plane displacement of the SMP skin enhanced with increasing temperature and the maximum out-of-plane displacement was decreased rapidly by using the pre-strain method. They also developed morphing wing structure using sandwiched skin technology [35]. SMP embedded with wire springs was used for morphing wing skin where heating wire springs acted as the activation system for the SMP.

5.5 Miscellaneous

SMPs have been used as packaging materials in the preparation of heat shrinkable tubes and films [2]. Upon heating the heat-shrinkable tubes at T_{trans} will come to their original shape and provide fixation or cover the objects. It is very much useful in cable industries. Heat-shrinkable films are particularly suitable for packaging a number of articles in a single container. The articles are placed in a bag or wrapped by heat-shrinkable film and then subjected to heating at T_{trans} to shrink the film around the articles. Single articles can then be removed separately while the remaining articles are kept together by the shrunk portion of the film. SMPs are suitable for packaging and depackaging of objects. In case of depackaging, no need of tearing off or cutting the packaging material just heating above the T_{trans}.

SMPs are gaining interest in automobile engineering. Many products have been developed such as reconfigurable storage bins, seat assemblies, tunable vehicle structures, hood assemblies, releasable fastener systems, airflow control system and morphable automotive body molding [3, 37]. Modulus of elasticity, shape, size and orientation of SMP-reconfigurable storage bin can be changed upon external activation and then different articles can be inserted as required. The energy-absorbing properties of the SMP-based hood assemblies can be changed by changing shape, dimension and/or flexural modulus property of SMP upon external activation. Airflow conditions of a vehicle have strong impact on vehicle performance such as vehicle drag, vehicle lift and downforce and cooling/heat exchange. Reductions in vehicle drag enhance fuel economy. Vehicle lift and downforce can affect vehicle stability and handling. SMP-based airflow control system will be suitable as it can respond to changes in the environment.

SMPs can be used as active assembly/disassembly, i.e., components/parts are firmly placed together or separated without physically touching [38]. High recoverable strain, low cost and nonconductive nature enable SMP as a prominent material for active assembly/disassembly system. SMP screw is commonly used as a part for active assembly. An SMP screw with/without thread can be utilized as a one-for-all solution for a range of various sized holes and even a screw driver is not required for tightening.

Self-peeling reversible adhesives can be made by using SMP. Xie and Xiao [39] developed epoxy-based SMP self-peeling reversible dry adhesive. They prepared this reversible adhesive through combination of an SMP and pressure-sensitive adhesive which showed features similar to those of a gecko foot.

SMP can be used in a spoon or fork handle for a disabled person incapable to grip. The handles are formed by fitting to a hand above T_{trans}, then the shape becomes fixed below T_{trans}.

Smart mandrels can be made by SMPs. After the filament winding, the mandrel can be simply removed through the thermally induced shrinkage.

SMPs have been used as anticounterfeit stickers. Nanoshel company, India, has developed SMP-based anticounterfeit tags and labels, and these have higher security level than the ordinary tamper evidence label.

References

[1] Small IV W, Singhal P, Wilson TS, Maitland DJ. Biomedical applications of thermally activated shape memory polymers. J Mater Chem 2010;20(17):3356.

[2] Hu J, Zhu Y, Huang H, Lu J. Recent advances in shape-memory polymers: structure, mechanism, functionality, modeling and applications. Prog Polym Sci 2012;37(12):1720–63.

[3] Leng J, Lan X, Liu Y, Du S. Shape-memory polymers and their composites: stimulus methods and applications. Prog Mater Sci 2011;56(7):1077–135.

[4] Mather PT, Luo X, Rousseau IA. Shape memory polymer research. Ann Rev Mater Res 2009;39(1):445–71.

[5] Meng H, Li G. A review of stimuli-responsive shape memory polymer composites. Polymer 2013;54(9):2199–221.

[6] Maitland DJ, Metzeger MF, Schumann D, Lee A, Wilson TS. Photothermal properties of shape memory polymer micro actuators for treating stroke. Lasers Surg Med 2004;30(1):1–11.

[7] Small IV W, Metzger MF, Wilson TS, Maitland DJ. Laser-activated shape memory polymer microactuator for thrombus removal following ischemic stroke: preliminary in vitro analysis. IEEE J Sel Top Quantum Electron 2005;11(4):892–901.

[8] Small IV W, Wilson TS, Buckley PR, Benett WJ, Loge JM, Hartman J, et al. Prototype fabrication and preliminary in vitro testing of a shape memory endovascular thrombectomy device. IEEE Trans Biomed Eng 2007;54(9):1657–66.

[9] Baer GM, Small W, Wilson TS, Benett WJ, Matthews DL, Hartman J, et al. Fabrication and in vitro deployment of a laser-activated shape memory polymer vascular stent. Biomed Eng 2007;6(1):43.
[10] Kim JH, Kang TJ, Yu WR. Simulation of mechanical behavior of temperature-responsive braided stents made of shape memory polyurethanes. J Biomech 2010;43(4):632–43.
[11] Chen MC, Chang YC, Liu CT, Lai WY, Peng SF, Hung YW, et al. The characteristics and in vivo suppression of neointimal formation with sirolimus-eluting polymeric stents. Biomaterials 2009;30(1):79–88.
[12] Tamai H, Igaki K, Kyo E, Kosuga K, Kawashima A, Matsui S, et al. Initial and 6-month results of biodegradable poly-l-lactic acid coronary stents in humans. Circulation 2000;102(4):399–404.
[13] Venkatraman SS, Tan LP, Joso JFD, Boey YCF, Wang X. Biodegradable stents with elastic memory. Biomaterials 2006;27(8):1573–8.
[14] Yang CS, Wu HC, Sun JS, Hsiao HM, Wang TW. Thermo-induced shape-memory PEG-PCL copolymer as a dual-drug-eluting biodegradable stent. ACS Appl Mater Interfaces 2013;5(21):10985–94.
[15] Nakasima A, Hu JR, Ichinose M, Shimada H. Potential application of shape memory plastic as elastic material in clinical orthodontics. Eur J Orthod 1991;13(3):179–86.
[16] Jung YC, Cho JW. Application of shape memory polyurethane in orthodontic. J Mater Sci Mater Med 2010;21(10):2881–6.
[17] Lendlein A. Biodegradable, elastic shape-memory polymers for potential biomedical applications. Science 2002;296(5573):1673–6.
[18] Ortega JM, Small W, Wilson TS, Benett WJ, Loge JM, Maitland DJ. A shape memory polymer dialysis needle adapter for the reduction of hemodynamic stress within arteriovenous grafts. IEEE Trans Biomed Eng 2007;54(9):1722–4.
[19] Hu J, Chen S. A review of actively moving polymers in textile applications. J Mater Chem 2010;20(17):3346.
[20] Boczkowska A, Leonowicz M. Intelligent materials for intelligent textiles. Fibres Text East Eur 2006;14(5):13–7.
[21] Gök MO, Bilir MZ, Gürcüm BH. Shape-memory applications in textile design. Procedia Soc Behav Sci 2015;195:2160–9.
[22] Hu JL, Dong ZE, Liu Y, Liu YJ. The investigation about the shape memory behavior of wool. Adv Sci Technol 2008;60:1–10.
[23] Ding XM, Hu JL, Tao XM, Wang ZF, Wang B. Free volume and water vapor transport properties of temperature-sensitive polyurethanes. J Polym Sci Part B Polym Phys 2005;43(14):1865–72.
[24] Jeong HM, Ahn BK, Cho SM, Kim BK. Water vapor permeability of shape memory polyurethane with amorphous reversible phase. J Polym Sci Part B Polym Phys 2000;38(23):3009–17.
[25] Hu J, Meng H, Li G, Ibekwe SI. A review of stimuli-responsive polymers for smart textile applications. Smart Mater Struct 2012;21(5):053001–24.
[26] Meng Q, Hu J, Zhu Y, Lu J, Liu B. Biological evaluations of a smart shape memory fabric. Text Res J 2009;79(16):1522–33.
[27] Hu J, Zeng F, Li P. Methods of manufacturing deodorants, and deodorants resulting thereof. US 2005/0232880 A1.
[28] Liu Y, Du H, Liu L, Leng J. Shape memory polymers and their composites in aerospace applications: a review. Smart Mater Struct 2014;23(2):23001–23.
[29] Hager MD, Bode S, Weber C, Schubert US. Shape memory polymers: past, present and future developments. Prog Polym Sci 2015;49–50:3–33.
[30] Lan X, Liu Y, Lv H, Wang X, Leng J, Du S. Fiber reinforced shape-memory polymer composite and its application in a deployable hinge. Smart Mater Struct 2009;18(2): 024002–8.
[31] Barrett R, Taylor R, Keller PN, Lake MS, Stern T, Freebury G, et al. Design of a solar array to meet the standard bus specification for operation responsive space. ASC Struct Struct Dynam Mater Conf 2007;(April):1–11.

[32] Zhang R, Guo X, Liu Y, Leng J. Theoretical analysis and experiments of a space deployable truss structure. Compos Struct 2014;112(1):226–30.
[33] Barrett R, Taylor R, Keller P, Codell D, Adams L. Deployable reflectors for small satellites. 21st Ann AIAA/USU Conf Small Satell 2007;(303):1–6.
[34] Yin W, Liu J, Leng J. Deformation analysis of shape memory polymer for morphing wing skin under airflow. Front Mech Eng China 2009;4(4):447–9.
[35] Yin WL, Sun QJ, Zhang B, Liu JC, Leng JS. Seamless morphing wing with SMP skin. Adv Mater Res 2008;47–50:97–100.
[36] Yu K, Yin W, Sun S, Liu Y, Leng J. Design and analysis of morphing wing based on SMP composite. Ind Commer Appl Smart Struct Technol 2009;7290(2):72900S–72900S–8.
[37] Leng J, Lu H, Liu Y, Huang WM, Du S. Shape-memory polymers – a class of novel smart materials. MRS Bull 2009;34(11):848–55.
[38] Sun L, Huang WM, Lu HB, Wang CC, Zhang JL. Shape memory technology for active assembly/disassembly: fundamentals, techniques and example applications. Assem Autom 2014;34(1):78–93.
[39] Xie T, Xiao X. Self-peeling reversible dry adhesive system. Chem Mater 2008;20(9):2866–8.

6 Future directions

The research on shape memory polymers (SMPs) has resulted that they have wide range of applications in numerous fields. Various applications of SMPs have been explored and some of them have been commercialized. Though many exciting achievements have been made, still some challenges have to be overcome for ideal application of SMP. The current development and application of SMP in fact lag behind other smart materials. The synthesis of novel hyperbranched, interpenetrating network and blend SMP would definitely overcome the existing downsides. In addition, development of biobased SMP will be a fortune for environmental-friendly and biocompatible SMP. SMP nanocomposites have played an important role to improve the shape memory properties of SMP. However, detailed understanding of the effect of nanomaterial on the shape memory effect (SME) will render wonderful SMP. The appropriate classification of SMP would play an important role in interpreting the underlying mechanisms of the SME, and subsequently in understanding the design of new SMP. Remote-actuated SMPs such as electroresponsive and magnetoresponsive have been developed. However, the required high actuation power and low recovery force limit their applications. Suitable remote-actuated SMPs are expected to become one of the leading SMPs in near future. A comprehensive characterization of SMP is crucial to understand the relationship between structure and SME of SMP. However, advances in existing characterization techniques and development of new techniques will offer the foreseeable future. Computational modeling and simulation may also play a unique role in the prospect of SMP characterization, essentially in initial feasibility and proof-of-concept studies. Durable, excellent recovery behavior, excellent biodegradable and biocompatible SMPs are extremely needed for biomedical applications. Suitable mechanical properties of SMPs can be achieved for each individual biomedical application by the in-depth analysis of structure–property relationships of SMP. Moreover, excellent thermostability, high modulus and excellent recovery stress are the prerequisites for aerospace and automotive applications. Again precise multiresponsive SMP would significantly widen the engineering applications of SMP in future. Extensive and intensive studies on the multiresponsive SME definitely open the new door of SMPs. In the last decade, it was seen that a large number of patents related to SMP products were filed. It can be predicted that in the next couple of years, many more SMP products will be realized and be commercialized.

https://doi.org/10.1515/9783110570175-006

6 Future directions

Index

https://doi.org/10.1515/9783110570175-007